The Stranger behind

the

Copernican Revolution

Notes, letters and diary entries
of

G. Joachim Rheticus
1539-1543

The Stranger behind

the

Copernican Revolution

Notes, letters and diary entries
of

G. Joachim Rheticus
1539-1543

Ulrich Maché

The Stranger behind the Copernican Revolution
Copyright © 2005 Ulrich Maché
Published by Hudson Books

All rights reserved. No part of this book may be reproduced (except for inclusion in reviews), disseminated or utilized in any form or by any means, electronic or mechanical, including photocopying, recording, or in any information storage and retrieval system, or the Internet/World Wide Web without written permission from the author or publisher.

Cover art courtesy of NASA Jet Propulsion Laboratory,
California Institute of Technology

Book design by:
Arbor Books
www.arborbooks.com

For further information, please contact:
info@arborbooks.com
or
1-877-822-2500

Printed in United States

Ulrich Maché
The Stranger behind the Copernican Revolution
Library of Congress Control Number: 2005930370
ISBN:0-9767789-6-3

Acknowledgements

Of the 757 books and articles on Copernicus which the *Library of Congress Online Catalog* lists today, I am indebted to a greater number than the general reader would want to know. Nevertheless, I wish to express my gratitude to all I consulted. My special thanks, however, go to Leopold Prowe and Karl Heinz Burmeister, the authors of the most comprehensive studies of the lives of Nicolaus Copernicus and Georg Joachim Rheticus, the two protagonists of this book. Prowe with his voluminous study on Copernicus, and Burmeister with three volumes on Rheticus opened up a new world for me, bringing to life the extraordinary circumstances that led to the publication of a tome which changed our concept of the universe. Furthermore, I am indebted to Edmund Robertson and Thomas Minnes for providing some of the illustrations for this book, and would like to thank my wife Britta for her unflinching support.

Contents

Biographical Background 1

Hope and Frustration. 5

Toasting the Astronomers 45

Eureka. 75

Biographical Comment. 125

Epilogue 131

Historical Figures:
Biographical Highlights. 135

Biographical Background to the Rheticus Papers

Nicolaus Copernicus (1473-1543) would never have witnessed the publication of his momentous discovery, had not Joachim Rheticus, a most promising mathematician from Luther's University at Wittenberg, taken the initiative in 1539 and urged the aging astronomer to prepare the manuscript for the press. At his own expense Rheticus made the arduous journey from Saxony to Varmia in northern Poland, stayed with Copernicus for more than two years and, in the summer of 1541, spent six exhausting weeks copying the manuscript he helped revise, and carried it himself to the printer in Nuremberg.

Prior to Rheticus' arrival in Varmia, Copernicus had done very little to disseminate his mind-boggling idea of a sun-centered universe. Pressed hard by his friends he had produced a sketchy draft of it entitled *Commentariolus*. It had circulated among scientists for more than twenty-five years, but without any impact. Finally, in 1537 the script caught the attention of Joachim Rheticus, whose genius immediately sensed the scientific significance of Copernicus' idea, but was puzzled by the lack of convincing data and the vagueness of certain arguments. Exceptionally gifted and a professor of mathematics and astronomy at the age of twenty-two, he decided to discuss the critical issues with a number of distinguished colleagues. For this reason he traveled to Nuremberg to visit Johannes Schöner, a celebrated astronomer and mathematician, and subsequently to Tübingen, where he met the renowned Joachim Camerarius. In certain ways, both trips were a success: Schöner took to the young man immediately, and their relationship grew into a lasting friendship; Camerarius was impressed with Rheticus' genius and learning and recommended him a few years later for a position at the University of Leipzig. Yet none of these luminaries could satisfy Rheticus' questioning mind.

So he decided to see the controversial astronomer in person. With a leave of absence from Wittenberg University for the spring of 1539, he set out for the Baltic Sea. Accompanied by his assistant, Heinrich Zell, who was to work independently on a map of the Prussian coastline, he reached the delta of the Vistula River in early May. While Heinrich Zell began his survey near Danzig, Rheticus headed toward Elbing and soon reached Frauenburg, that tiny town on the Vistula Lagoon with its oversize cathedral where Copernicus had spent most of his life administrating rural estates belonging to the diocese of Varmia.

The Stranger behind the Copernican Revolution

The "Rheticus papers" published here for the first time are fictitious in nature but adhere closely to established biographical facts and historical happenings. Readers not familiar with the age of Humanism and Reformation will find the index of historical figures (p. 135-142) helpful.

Northern Germany and Poland 1539

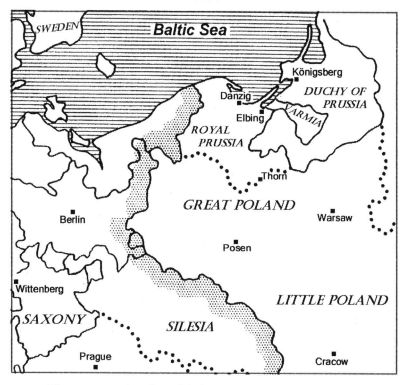

The western border of Poland separated that country from the "Holy Roman Empire [of the German Nation]" (962-1806)

I

HOPE

AND

FRUSTRATION

Frauenburg

Seventeenth century rendering of the town. In the foreground: the waters of the Vistula Lagoon (Frisch Haff)

***
Diary Entry
Frauenburg, May 7, 1539

The young woman who opened the door was unusual — so self-assured, rather urbane, something one would not expect in this remote corner of the world. When she smiled and said, " Dr. Nicolaus has been expecting you", I knew that my letter from Posen had reached him.

These canons of the Catholic Church seem to lead a rather worldly existence. I had expected a kind of monastic atmosphere, but *NO!* In addition to their lodgings at the Cathedral they have their own mansions out of town with stables, servants and housekeepers. To my amazement they live as well or better than any patrician in Wittenberg or Nuremberg. My little guest room on the upper floor is not exactly what I am used to. It's certainly drab compared to the living quarters downstairs, but I am not complaining. I consider myself lucky to be here, to be close to this unusual man, and in a few days I'll find some suitable quarters in town.

***
To his Mother,
Thomasina de Porris in Bregenz
Frauenburg, June 2, 1539

You will be relieved to hear that your son Joachim finally made it to Frauenburg. The place is even smaller than I thought, not exactly your type of town. Coffee is unavailable here, but I'll survive, and so would you. If you wanted to find Frauenburg on a map, you would have to crawl into the southeast corner of the Baltic Sea.* I am hoping to stay here for some time, and I expect to be very busy. If you don't hear from me, you have to take it as a the best of signs. And above all, don't worry about your Lutheran son in a Catholic country! Of course, it *can* be dangerous; but don't forget: I am a diplomat. You can rest assured that I will not end my life like my father. By the way, I must thank you again for being so protective of me when it all happened. You made very little of his terrible fate, torture and all. In retrospect I see that you did the best thing you could have done. I was too young to grasp what was really going on. I somehow knew, and yet, I did not really know until I asked you for an explanation last year and you wrote down the whole story in that remarkable letter of yours. I cherish it and have read it many times with great pride. I even took it along on this trip, although I haven't looked at it since I read it to Schöner in Nuremberg. He knew about the case and was anxious to learn the details.

 The man I am seeing here is the most innovative thinker I know. You may not believe me, but I know that his ideas are bound to change the world. Hold your breath, mother, and put down your Bible. According to his calculations, the earth is moving around the sun. Everything in the heavens is different from what everybody had thought, and your little J. will be instrumental in spreading the

* See map Northern Poland 1539 on Page 43

unsettling news. It is so exciting that I would like to take a giant brush and paint in huge letters all over the sky, "Copernicus has seen the light! — Copernicus is right — is right, is right!". You may think I am out of my mind, but I am not — nor is *he*. And from what he tells me, all the confusion in our calendar will finally end. The Catholics will no longer fast on days when they could eat their mutton and their beef, and you will be able to celebrate Easter and Christmas on the right days!...

By the way, you would like him as a person. He shares with you a love of luxury. And he was used to it to an unusual degree. His uncle was Prince Bishop here in Varmia and for years Copernicus was, if I am not exaggerating, the strong man the throne. In any case, his days of glory ended with his uncle's death, two years before my birth. By rank, he is no more than a canon today, a humble canon, which strikes me as peculiar. If he had the ambition of his uncle, he would be a bishop or cardinal by now. But from what I see, he does not quarrel with his fate. In any case, his love of luxury survived, and Sebastian, his valet, dotes on him. From what I heard, he is not on best terms with the bishop, his superior, who recently forced him to dismiss his housekeeper, a very capable woman who is said to have fallen for him. All this is completely new to me, the latest scoop that I picked up from one of his cronies, Alexander Scultetus*, who drank too much last night and let me in on more gossip than I can reveal to you in this letter. For the more juicy details of Dr. Alexander's tales you'll have to wait until I visit you.

As for money matters: Avoid your Bregenz lawyers.

* *Also Sculteti*

I am glad Dr. Anger considers your recent investments in Feldberg a "wise move", and I agree. It's the best way to keep your assets apart from those of your new husband. (Not that I have anything against him!) Once again, Dr. Anger is the best lawyer we have in Wittenberg, and I suggest that you stay with him.
Much love,
J.

***
To Johannes Schöner,
astronomer and geographer in Nuremberg
Frauenburg, June 7, 1539

I am sitting at a small table in a guest room in Copernicus' house; it's not his "Curia" by the Cathedral but a kind of country seat just outside of town — not a bad place to be; and I am, I think, the happiest man on earth! Dr. Nicolaus has been in a splendid mood from the day of my arrival. By the way, I left my assistant near Danzig. From there he is working toward Elbing on a project in which I lost interest: a detailed map of Prussia. As I came uninvited and as a stranger to Frauenburg, I thought it wiser to make a solo appearance. I can assure you, it was a success, and things look *very* promising. A good number of my crucial questions have been answered; but as is always the case when you delve deeper into a subject, things turn out to be more complicated than

The Stranger behind the Copernican Revolution   11

expected. — I am, by now, studying various parts of the script, and I must tell you: it *is* impressive. Soon I'll be reading the whole from beginning to end. Only one thing worries me. Dr. Nicolaus does not want to discuss any plans for the publication of his work. I told him about your willingness to secure a printer, but he insists that the work is not ready and invariably changes the topic. I don't understand this because the work *is* virtually finished. After a few revisions and corrections he should be able to part with it before the end of the year. Even Dantiscus, his bishop, has urged him to publish. If he remains adamant, I might simply prepare a readable summary for publication. I strongly feel we ought to let the world know what this genius has done. And yet, how can I do that against his will?

You might be interested to know that Dr. Nicolaus is familiar with all your publications — and very interested in meeting you personally. He asked me to report at length about the time I spent with you in Nuremberg and your work on Regiomontanus. Curiously enough, he also wants to know everything and anything connected with student life in Wittenberg, and most of all, he loves to talk about his often boisterous student days in Italy.

Your suggestion to collect data on Copernicus' life while I am in Varmia appeals to me more and more. I have made a vow to keep a record of all the happenings during this visit, perhaps in form of a diary, but I am not sure that I can live up to my promise. In any case, I am going to save the drafts of the letters I'll be writing to

various people. Before long, you'll hear from me again. I have not forgotten my promise to keep you posted.
Joachim.

***

Diary Entry:
Frauenburg, June 19, 1539

Fog ... fog ... fog .... every day I wake up to dense fog, and only gradually — in the course of the morning — the Lagoon unveils itself grudgingly. Everybody tells me how unusual the fog is at this time of the year, but I don't believe it. On the other hand, things have been surprisingly good at night. That does, of course, not change the fact that, by and large, this place is a nightmare for any serious observer of the sky. Even my Mentor admits to that. And yet, here he made it happen! Here he stood many a night — watching, measuring, thinking — and, in the end, achieving the incredible: contradicting the ancients, shattering Ptolemy's beautiful universe. But *HOW?* How did he do it? How did it happen? Yesterday when I tried to get an answer, he was strangely evasive. I have to approach the subject from another side. There is nothing in this world that I am more anxious to know than the circumstances and conditions that led to the point where he became convinced that — contrary to all evidence — our earth is behaving like a planet, moving like a heavenly body around the sun. To know what went on in his mind at those decisive moments, I could sell my soul. — God forbid!

The Stranger behind the Copernican Revolution

***
To Johannes Schöner,
astronomer and geographer in Nuremberg
Frauenburg, June 21, 1539

You wanted a description of his "observatory". There is no such place; nothing like the one Regiomontanus built in Nuremberg. What my Mentor has here is a kind of platform on top of a wall close to the tower that he has occupied for many years and that he still uses from time to time. For observations this platform is quite convenient, but it is nothing more than an open terrace. — You should see his collection of instruments: by Nuremberg standards they are primitive, shockingly primitive; some of them home-made! Can you imagine that this genius, this second Ptolemy never bought himself an armillary sphere? Not that he was ever short of funds to buy the best! As I am told, he is loaded with money—and not stingy either. From what I see, he knows how to live—and does so almost as well as my mother!

Yes, the lifestyle of these canons at Frauenburg is rather aristocratic. Their duties are minimal, and their incomes from Church estates considerable. Picture a university professor at Paris or Bologna, lavishly paid, with no obligation to teach, no pressure to publish, and you get the idea ... I wish I could show you the external setup at this cathedral. It's very curious indeed, but I like it. Unbelievable, these buildings and fortifications surrounding the church: bastions and walls that are almost as massive and high as the ones at Nuremberg.

However, they were not built to protect the town but the church. I haven't seen anything like it in all of Germany. Alexander Scultetus, who loves wine as much as I do, often jokes about this and calls it "the fortress of God" because in case of war the canons leave the defense to Him. At the outbreak of hostilities in 1520 they disappeared overnight, running for shelter to cities like Elbing and Danzig.

To me, who grew up in the south where Christianity was established more than a thousand years ago, it sounds strange that the Cross is so new to this region, hardly more than two hundred years. If you believe Scultetus, the hilltop on which the cathedral now stands has been the favorite meeting place for wolves since the time of Noah's Ark. Here they convened nightly "to wail and yelp at the moon" until in 1329 a Christian bishop put an end to their ungodly yowls by crowding out the Prussian wolves by "a pack of German priests".

— So much from the grapevine of Dr. Scultetus, who in his sober hours is working on a *Chronology of the World*, in which he will — in all seriousness — immortalize Dr. Nicolaus as "the Ptolemy of our millennium".

The Stranger behind the Copernican Revolution

***

To his friend and colleague Paul Eber
in Wittenberg
Frauenburg, June 22, 1539

 Your letter arrived today, and I am trying to answer your administrative questions as briefly as I can; after that, a little bit of dirty laundry from Varmia. To be frank with you, I am unwilling to return to Wittenberg for the winter semester. In other words, do not schedule me for any courses beginning in October. I would, of course, like to keep Heinrich Zell up here. He has done excellent work on the map of Prussia and I hope he will be permitted to finish it. At the moment he is incorporating geographical data which Copernicus and his friend Scultetus have gathered for a similar project many years ago. As Heinrich's chief interest has always been cartography, he just loves this project. Should the university want him back for the fall, I seem to have no choice but to let him go. However, Melanchthon should realize that an accurate map of the coastline would enhance Wittenberg's reputation, and Luther ought to be reminded that the map would be a splendid gift for Duke Albrecht of Prussia, the only defender of Protestantism in this remote region. Please, see what you can do for me at the upcoming committee meeting.
 — My Mentor has had some correspondence with the Duke, and I hope to see him in person when I get to Königsberg. But when?
 For an extension of my leave, I'll write, of course, to

Philipp Melanchthon
On religious grounds he rejected Copernicus' theory, but fully endorsed Rheticus' scientific ventures. - - - Exceptionally gifted himself, Melanchthon entered the University of Heidelberg at the age of twelve and received his B.A. two years later. His educational reforms earned him the epithet "Preceptor of Germany".

Melanchthon, hoping to find him in a generous mood. He has always been good to me, and I think he has forgiven me by now for not taking a wife after my appointment to the faculty. To boost his magnanimity I intend to send him the draft of an essay that shows the compatibility of Copernicus' theory with the Holy Scriptures. If that should not lead to the granting of a sabbatical, I

The Stranger behind the Copernican Revolution

have decided to disregard the University's decision. I shall stay here regardless of the consequences. No word of this to anyone please!

You asked about the rumored manuscript *On the Revolution of the Heavenly Spheres* . . . It is sitting on the shelf in front of me — five hundred and thirty-six pages in its present state. I have begun studying it and intend to work through it page by page. That takes time, but it is the most fascinating reading I have ever done. And it is a wonderful feeling to have access to the author who is willing and eager to clarify things for me. To my surprise he is even anxious to revise some murky spots that bother me. This open-mindedness, this lack of vanity is so refreshing, something rarely found among our beloved colleagues at Wittenberg.

My old prediction stands: the book is bound to make an enormous splash. — I can't wait! The trouble is that my Mentor is not ready and willing to publish it. For me, who cannot see things in print fast enough, this is a little hard to swallow. Last night I had a dream in which I secretly stole the script and took it to Mainz for publication. In the dream the manuscript was hardbound and heavier than Gutenberg's bible. I was just handing it to the printer when my Mentor appeared and jerked the book out of my hands lifting it up high and with a powerful blow hitting me over the head. At that point I woke up and realized it was a dream. As you can see, I have a problem, but I am sure things will work out.

I know you'll love this bit of gossip: Upon the Bishop's repeated behest my Mentor fired his housekeeper last December! "New Morals for Varmia!", says Scultetus, my drinking companion, "It's Dantiscus riding his hobbyhorse." From others I hear that my Mentor's housekeeper was an "accessible" woman (whatever that means) with a breath-taking cleavage which caused rumors to spread! God knows how much of this is exaggerated, but life in Varmia is certainly not what it used to be before Dantiscus' arrival two years ago. As they say here, the new Bishop has a hundred eyes, and as a result Scultetus is in deeper trouble than I ever imagined. Yet, as a former emissary to the Vatican and a historian of note, this amiable fellow has a self-confidence to be reckoned with. I venture to say that he is afraid of nothing. Relying on his old friends in the Papal Curia, he is now taking his case to Rome. The charges brought against him range from heresy to — hold your breath — "illicit penetration". As you see, Virtue has invaded this remote corner of the globe. The irony of it all is that Dantiscus himself is a man with a lustful past: his volume of libidinous poetry, which I devoured as a teenager, leaves nothing unsaid, and during his diplomatic career he is said to have left a fortune in the brothels of Vienna and Paris. Add to this that in Madrid he embarrassed the court with a theater sketch "in the Roman tradition" and you get the picture. Magically his past became a white cloud in the sky when his Excellency ascended to episcopal glory and ushered in what Scultetus wryly calls "the age of fornicatory demise in Varmia."

    I am almost certain that the allegations against my

The Stranger behind the Copernican Revolution 19

Mentor are largely unfounded. He is now reproaching himself for having yielded too easily to episcopal pressure. At present he is visiting Danzig, and I do not expect him back until next week. Before his departure he passed by, and from his remarks I gather that he is also meeting that woman. — You should have seen him departing in style, accompanied by two servants. Impressive!

All along, he has been a most gracious host, and before he left he sprang a surprise that I found touching and overwhelming. For the length of my stay at Frauenburg he permitted me to use his old study, located in the fortifications surrounding the cathedral. The last

Frauenburg Fortifications
with the so-called "Copernicus Tower". In the distance: the Vistula Lagoon (Frisch Haff) with its sand spit on the horizon.

ten years he has worked and lived primarily at his "extramural home", a kind of lordly estate not far from town. That is, of course, more convenient for him, and the place is much more comfortable. But I love these old fortifications and, above all, the tower where he did most of his work. His books on astronomy are all here, and, better still, there is the desk at which he wrote his book! — I just love the thought that his desk is now my workplace — my place of work! At times I get so excited over the idea that I have to pound my fists on it to regain a sense of reality. Oh, Paul, I wish you were here at such moments, I wish you could see it all. I wish I could embrace you! Life is great, and the joy to live is tremendous!

Your Joachim

***

Diary Entry:
Frauenburg, June 25, 1539

    I have worked incredibly hard on Book II of my Mentor's manuscript, reading all day and staying up late to verify certain measurements on Mercury. It's been exhausting, very exhausting! How long can one sustain this type of exertion? — Everything has its price: when he returns from Danzig, I want him to be impressed with what I have done. I hope he'll be able to get some coffee for me. Do I ever love that Muslim brew!

The Stranger behind the Copernican Revolution

***

To Johannes Schöner,
astronomer and geographer in Nuremberg
Frauenburg, June 27, 1539

Big news from Wittenberg: Luther is objecting to my stay in Varmia. According to the latest scoop, Rheticus "is wasting his time with a loony astronomer and a gang of shady papists". Dr. Martinus seems to be as hotheaded as ever. In a committee meeting he slammed his hands on the table, yelling that he would never accept the notion of a moving earth. Fortunately, Paul Eber diverted him from the issue, insisting that my work up here with Heinrich Zell was — to a large degree — "a cartographic research mission", much in the interest of Duke Albrecht of Prussia and hence a promotion of the Protestant cause.

You will be delighted to hear that Dr. Nicolaus finally removed my last doubts about a calendar reform. With the adjustments he has in mind, all the incongruities that have crept in since Ptolemy's time can be swept aside. The only drastic measure would, of course, involve advancing the calendar by about ten days. From then on until doomsday, Catholics will be able to feast and fast in sync with Copernicus' heavenly spheres! My Mentor told me jokingly that his old buddy, Tiedemann Giese, who is now Bishop of Culm, jokingly predicted that a successful calendar reform should earn him exemption from purgatory and instant promotion to sainthood.

Unfortunately nothing has changed in his attitude toward publishing. He still becomes tense, even feisty at any suggestion to go public with his theory. The mere prospect of putting his cards on the table seems to frighten him.

When I first noticed this, I assumed that it was a kind of shyness or reluctance that would soon be overcome. By now I know that it is a more serious problem, something he cannot shake off easily, and I do not see an easy way out. On the other hand I cannot stand by idly and wait. In fact, I have started to work on a summary of his book. It ought to be published as soon as possible and should pave the way for the appearance of his *magnum opus*. If he should object to the printing of my summary, I might go ahead without his consent. There is a voice in me that says, "He has no right to withhold the truth with which God has inspired him. The world ought to know!" Yes, I hear that voice time and again!

***

Diary Entry:
Frauenburg, July 3, 1539

Rain and more rain. Yet he did return! I was delighted and flattered that he would drop in on his way home. He and his servants were drenched. — I could tell immediately that Danzig had been a disappointment. When he entered the room, he looked

The Stranger behind the Copernican Revolution

paler than ever, careworn and haggard. In fact, he appeared outright depressed. Nevertheless, he claimed to have had "a splendid time", repeating that twice! While Sebastian went down to the horses to get my coffee, I was stupid enough to talk about the work I have done. He pretended some interest, but I could see that nothing registered with him. Is it this woman? This Anna Schillings? His mind was far away.

***

Diary Entry:
Frauenburg, July 7, 1539

Today he sent Sebastian to find out whether I needed anything. That has not happened before. As a rule, he would pass by in the afternoon to talk for a while, usually before going to mass. Sebastian is still unmarried at thirty-five and very good-looking, and I love to listen to his West-Prussian accent. — He told me that, since their return from Danzig, his master had spent most of his time in bed, but denied that Dr. Nicolaus was sick. "Not sick. Just resting; resting and reading and . . . moping a little. That's all, just moping. Nothing to worry about. He'll be on his feet in no time because he's back on his magic potion." I looked at him questioningly and smiling. "Yes", he went on, "aqua regia. The Doctor makes it himself, and believe me, I swear by the stuff. Two years ago it saved my life, and the bishop has tried it too . . . most of the canons as well. — One of them died, but I think he would have

Nicolaus Copernicus
Late sixteenth century woodcut. The lily of the valley identifies Copernicus as a medical expert. – Although he studied medicine in Italy, he did not take a medical degree. Nevertheless, he became his uncle's personal physician and acquired a legendary reputation as a "doctor" among the peasant population of Varmia.

died anyway. — You probably know that the Doctor used to have quite a lab in the basement of this tower. He's moved everything to the mansion now." I pretended I knew, and volunteered the information that, after all, his master had studied some medicine in Italy.

Sebastian was not impressed. "Before my time", he continued, "my master had quite a reputation as a doctor. He even worked miracles among the peasants, but he stopped that; he's getting a little old. It's just friends now and family — the canons and the bishop and myself. I always carry a small bottle with me." Pounding his chest he hit his supply of aqua regia, and reaching inside his coat, he pulled out a flat flask. — I thought I had him in a mood in which I could extract from him some news about their Danzig visit, but to my surprise there was a stern resistance. He volunteered a three day stay with the Schillings and a few nights with the Gieses, but nothing more — except the old cliché of the doctor having had a "splendid time". —- Strange, very strange!

***

Diary Entry:
Frauenburg, July 12, 1539

Pleasant dream about Sebastian. —- Late this afternoon Copernicus came to see what I had done in the last two weeks. We discussed most of the problems that had arisen and we both, I think, had a wonderful session. His eyes light up whenever I ask a question, and while he is thinking I seem to see his mind at work. It's an experience! His smiles and approving nods were a terrific boost to my ego. How I needed this after all that work! And yet, I feel guilty: I wanted to tell him the truth, but in the end I left him with the impression that all my notes were intended for lectures at Wittenberg. It bothers me more and more that I might be forced to publish the summary without his permission. Last night I woke up and could not get to sleep again. I walked around and stood outside on the platform, looking up into the starred heavens to those celestial spheres, thinking of him and his work. The wind came from the sea and I could hear the lapping of the waves on the shore of the Lagoon. Time and again I tried to understand him, but I always ended up with the question: "Isn't there something unethical about his reluctance to share his findings?" What can I do to help him?

What would happen, if I confronted him? — If he had to fear prosecution, I would understand. If there were even the slightest danger of it, I might see his point. But in fact there is no threat, and the logic of his behavior escapes me. What I would give to fathom that man!

The Stranger behind the Copernican Revolution 27

***
To Johannes Schöner,
astronomer and geographer in Nuremberg
Frauenburg, July 15, 1539

    This morning I took a walk with Dr. Nicolaus along the shore of the Lagoon. That body of water is remarkable. On overcast days it can give you the illusion of being the sea itself. Those are the days when my Mentor loves to walk along the edge of the water and get his boots caught by the waves. During such strolls he likes to reminisce about his boyhood in Thorn, his student days at Cracow and the carefree years in Bologna.
    I did not dare to touch the topic of publishing today, but the old problem persists: it's tormenting me more and more. I wish you could help. I am trying to be patient, still hoping to make him see that he is not facing a political fight — and certainly not a religious one, no matter what Dr. Martinus says. Nothing in Copernicus' theory contradicts the Scriptures: my defense (which he has read and of which I am enclosing a copy for you) has made that clear. And who is Luther when it comes to geometry, mathematics and astronomy? He can rave and rant and pound his belly, but he has no means to threaten my Mentor. Even if Copernicus lived in Wittenberg, Dr. Martinus could do nothing. There will, of course, always be squeals from dim-witted colleagues, but who is afraid of *them*? As you know, Rome has been clamoring for a calendar

Bologna

Around 1500 one of Europe's leading universities. Here Copernicus studied canon law, attended the astronomy lectures of Dominico di Novara and may, at one time, have boarded at Novara's house.

reform for almost a century. Dantiscus is pushing him, and so is the Bishop of Culm. Even one of the Roman Cardinals has encouraged him to come out. Dr. Nicolaus showed me the letter, but he wants no publicity, no scientific squabbles. I am helpless, and believe me, this helplessness is the most frustrating experience of my life! At night I wake up, unable to fall asleep again. It all seems so absurd; and yet, in every other respect this man is — as far as I can see — the most reasonable person on earth!

I promise you, I will not give up. I'll prepare his *De Revolutionibus* for the printer, hoping that in the end he will be ready to share his knowledge with the world.

Joachim

***

To his Mother,
Thomasina de Porris in Bregenz
Elbing, July 18, 1539

Surprise, surprise! I landed in what you might call a civilized town. It's not Danzig, the big city where I would love to go as soon as I can, but the seaport of Elbing. In this place, just twenty miles south of Frauenburg, business is booming. Everything here seems bigger than in Wittenberg — and certainly more affluent. You should see some of the houses and the inn where I am staying! Not large, but quite elegant. They even serve such exotic things as coffee here, something unheard of in Frauenburg. I wish you could have been with me this afternoon when I took a stroll along the harbor and the center of town. It's Sunday, and women were showing off their dresses. You should have seen some of the fashion-mongers in their latest craze from Spain. — You'll be delighted to hear that much of the town turned Protestant in 1525. As your minister-friend Sebelius would say,

"This Elbing is a refreshing spring in a Catholic wasteland". But guess what your little J. did this morning? He refreshed and cleansed his soul in church, the first time since he left Wittenberg at the end of April. However, church service was not what brought me here; nor was it the lure of Elbing's demimonde, whose main "house" is extremely popular with the sailors. It's reputedly one of the poshest of such establishments between London and Stockholm. No, I was summoned here by court order because Heinrich Zell, my assistant, is in prison. I had never expected him to get into such trouble because in Wittenberg he hardly drank. To give you his tale in a nutshell: While surveying the Baltic coastline for a map that we want to publish he was roped into town by an "itinerant scholar" and egged on by him to play braggadocio, moving from pub to pub, bragging and piling up debts and finally getting into a fight over a girl with two Danish sailors. They all ended up using knives with rather bloody results. While the "scholar" fled, Henricus was thrown into prison — not so much for mutilating one of the sailors as for his debts. I have visited him in his dungeon and should get him out after settling his accounts with the creditors.

Some day I expect to get my money back, but that is not an issue now. I'll let Henricus worry about that. As for myself, I want to return as fast as I can to my tower by the Cathedral. Yes, mother, I just love my work — this fascination and total absorption by the problems and mysteries of the heavens; it gives me incredible highs and makes life worth living. I sometimes ask myself, if I should write to you more about my work and my Mentor's momentous discovery, so that you can come to see that *not our earth* is the center of the universe —- *but the sun*. However, in the end I am always scared that I'd bore you with it. Yet, rest assured that any question you have about it will make me happy. Astronomy to me is *so* much more than the comforts and diversions of this city, much as I love them too! Saying this, I realize that my present stay in Varmia is only possible because I am blessed with a most generous and fortunately not impecunious mother. — Why do I always fall back on my training in oratory when it comes to writing an innocent little thank-you note to you? I know the answer, of course: you relish these touches of ancient bombast; and why should you not love to be charmed and cheered with a laudation that you deserve more than any mother on earth? So let me say loud and clear: Long live my gracious, beautiful mother, willing (and fortunately able!) to support her son's celestial adventures along with his follies — a mother willing even to pull out of dungeons some wretched knife-swinging devotees of cartography, be it at Wittenberg or on the remote shores of a northern lagoon! Oh, mother, let me embrace you, let me kiss you and thank you a thousand times!

Your little J.

P.S. I am still mulling over your wish to see Virginia as my wife. She is lovely, no doubt, and all you say is reasonable. Unfortunately, I am not ready to confront that issue now. Bear with me, Mother! . . . You'll hear from me.

***
Diary Entry:
Elbing, July 19, 1539

I am not giving up. When I get back to Frauenburg I have to talk to him once more. A new approach? But which? O God, will you give me the strength to confront him? I need your help! Your help! Your help . . .

***
Diary Entry:
Frauenburg, July 20, 1539

Scultetus told me again how Tiedemann Giese has tried for years to persuade Dr. N. to publish his book, and before that it was Giese again, who urged him to let the manuscript of *Commentariolus* make the rounds. — Scultetus' advice this morning was probably good: leave Dr. Nicolaus alone. — In spite of that I am determined to be courageous and force the issue. (O God, why do you make it so hard for me?)
Just before dark Sebastian came by to deliver a book that his master had wanted to bring himself. Unexpectedly, he was called out of town but will see me as soon as he returns. S. lingered; strange attraction.

The Stranger behind the Copernican Revolution

***

To his friend and colleague Paul Eber
in Wittenberg
Frauenburg, July 23, 1539

 I wish, dear Paul, that you were here to listen! I am desperate. Yes, desperate! And writing to you is just a try to let my emotions go and find myself again. Early this morning I had my first "encounter" with him. The *casus belli*, as you might guess: to publish or to perish with one's wisdom. I tried to say that he had an obligation to share with the world what — with the help of the Almighty — he has been able to discover. Although I could tell by his stare and the redness of his face, how much this infuriated him, I went right on with my arguments. I felt I had no choice. When I finished, an ominous silence filled the room. It seemed to last forever. Then he erupted. I have no other word to describe it. Like a volcano he erupted, covering me with ashes and cinders. I am still covered with them as I write this, shattered and devastated. Somehow I feel very guilty because too many times I had sensed his discomfort, and I knew that Giese and Scultetus had not been able to turn him around in all those years. Why should I succeed?
 Of course, I never expected this violent reaction. But looking back on the event, I can thank my creator that I am still sitting in this tower. Instead of dashing from the room and slamming the door, he could have thrown me out, could have asked me to leave

Frauenburg. In that case, the manuscript would sit on the shelf untouched to the day of his death and — under the best of circumstances — vanish in the archives of the Chapter. O Science, fair Lady, where art thou headed with this man! O Paul, fair Master of the Arts and of Theology, what comfort doest thou have to offer? I am at a complete loss.

A few days ago I returned from Elbing where Henricus Zell had been in trouble. Things have been straightened out, and he is back in the Delta, working his way toward Königsberg, where I hope to meet him when the time comes. — Please keep me posted on happenings in Wittenberg. On days like this your friendship means so much to me.

As always,
Joachim.

***

Diary Entry:
Frauenburg, July 24, 1539

    I did not sleep well, and I do not feel well. What is going to happen, if I get sick . . . under these circumstances?
    After lunch I went over to see Scultetus. I had to have somebody to talk to, some person to confide in. He was still sober and listened with great seriousness, shaking his head, or nodding from time to time. After I finished he looked at me with a sad smile, saying, "Let me talk to him. I am not sure what I can do, but I'll try. . . . I have known him for too long

to know that this is not his way of dealing with people. Give me a day, and in the meantime relax. — And take this as my advice: Stop pestering him with the specter of publishing, and even if he should mention his brother, don't ever ask any questions."
"A brother? I asked in surprise, because there had never been a hint that he had a brother.
"He was a canon here at the cathedral, an unfortunate case. In the end, they had to oust him from the Chapter because of his disease. He had contracted leprosy, but there is more to the story, but I should not talk about that, at least not today." Taking a hard look at me, he told me to my face that I didn't look well, that I ought to take better care of myself. I assured him that I felt as fit as ever. — Why didn't I tell him the truth?... Why this pretense? — Pride?

***

To Johannes Schöner,
astronomer and geographer in Nuremberg
Frauenburg, July 26, 1539

For more than a week, my dear Johannes, I had meant to write to you, but my days were full of inner turmoil and my nights filled with unrest. It all came to a head when I tried to press the issue of publication, and my Mentor, this most rational being, lost his patience with me and blew up. Scultetus, my historian and drinking companion tried to mediate. Since this morning things have changed for the better, and I am trying to recover from the strain. I assume that I have to thank Scultetus for intervening. Although I have not seen him for two days, I am convinced that his role in this development was decisive. My troubles started when I tried to corner my Mentor. Misjudging his dread of exposing himself to ridicule, I brought him to a point of uncontrollable rage. In a fierce fury he stormed out of the room, leaving me in a state of emotional hell. The last three days have been, I assure you, pure torture in the claws of Satan. By comparison, my recent trip to Elbing, which entailed a number of unpleasant dealings, seems like a carefree vacation. From the minute Dr. Nicolaus left me, time did not move: for days I mulled over what had happened and agonized over what I should have done; and all this the disturbing thought of whether or not a reconciliation was possible.

Fortunately, the solution came faster and less painful than I could have hoped. Last night, when I was still working in his study in the tower, Sebastian delivered a letter that I did not dare to open while he was present. After he had left I read it immediately and I was flooded with joy. I could not believe my eyes: Dr. N. even conceded that my prodding was "an unselfish act in the service of astronomy", that he appreciated having me as a guest and discussion partner, and that he hoped I would forgive him. In my excitement I threw up my arms as if I wanted to fly out of the room, but instead raced around the study like a madman. You should have seen me finally hopping out on the platform high above the wall of the fortifica-

The Stranger behind the Copernican Revolution 37

tion as if I were headed for a flight over the Lagoon. At that point, the memory of a conversation with Melanchthon brought me back to reality. I suddenly heard his voice, high-pitched and moralizing, "Beware of Icarus!", one of his favorite excursions into Greek mythology. So, instead of soaring over the Lagoon (or dropping to the ground below) I decided to take a deep breath and thank my creator for this incredible letter.

Unfortunately, I have not made much progress with the summary of Copernicus' book, but I have the feeling that the road toward publishing it has been cleared. — For some time I had meant to tell you that I intend to dedicate this slim volume to you, my dear Johannes. I trust you'll permit me this show of admiration for your edition of Regiomontanus and of my gratitude for what you have done for me at Nuremberg.

      As ever,
      Joachim

***
Diary Entry:
Frauenburg, July 26, 1539

This was a morning to be remembered. The visit of my Mentor was so extraordinary that I feel the need to spell it out in detail. I have to capture and preserve as much of it as I possibly can, both for my own sake, and for posterity.

He came to see me in his old study in the tower, the

room where I had cornered him, where he had left me covered with ashes! And yet, it felt as if nothing had ever happened between us. He was very excited about an emblem book that he had bought in Danzig, a slender volume, which he was going to leave with me for easy reading and entertainment, "a novelty, a sensation", he said.

To my surprise he wanted to hear about the progress I had made with my summary. I told him how I had labored to present in readable form the problems concerning Mercury and the section dealing with that planet's aphelion.

"I know exactly what you mean", he said, "I think nobody experienced that more painfully than I. In the beginning it was virtually impossible for me to prevent this interference of the orbital movements of the Ptolemaic system with my new concepts of our earth as a planet. Hundreds of times I tried to visualize this new universe; I tried to picture it with the help of drawings and sketches; I even built a crude model. I would never have thought that the workings of Ptolemy's magnificent inventions had been ingrained so deeply in my mind. But gradually I was able to shed much of my Ptolemaic past and was indeed able to visualize the movements and positions of the planets with a motionless sun at the center. However, to my surprise and frustration a big problem arose when I tried to couch all this in precise and persuasive language. I tried and tried, and finally gave up. For months I did not write a single line. And I must confess that I was simply scared to expose myself with this theory. I knew — and I still know at this moment — that most people would not accept it. Believe me, I went through a lot: the specter of ridicule and rejection just terrified me."

There was a long pause during which he rose from his chair and went to the window as if he were looking for something far out on the Lagoon or on the thin spit of the *Nehrung* on the horizon. "I just wanted to forget the whole thing; but that proved impossible! The idea of a moving earth just haunted me. So there was more thinking, more visualizing until I reached a point when everything seemed to fall into place. — What a relief! Then I started developing my own trigonometric procedures, and that was exhausting again, but great fun as well." He took a chair and moved it closer to me. "I know you wanted to talk about this before, but I just did not want to. I simply wasn't ready for it. I know it was stupid, but I had nothing against you. As you must have realized, and as I told you in my letter: I like you and I am glad you came here and are pushing me to put the last touches on the manuscript. I hope that during the next few weeks we will see a lot of each other." With these words he looked into my eyes while his left hand reached into a back pocket of his trousers pulling out a letter that he held up with a triumphant smile. "My old friend Tiedemann Giese hopes to see both of us soon. He knows about you and your reason for coming to Frauenburg; he expects us to stay with him at Löbau Castle for a few weeks. How about that? Would you be ready?"

Having heard enough about Tiedemann Giese from both Scultetus and my Mentor I readily accepted, saying I could go tomorrow if that was his plan.

"I do appreciate that", he said, getting up from his chair. "As you might remember, Dr. Tiedemann is the one who urged me to summarize my findings in the script that circulated as *Commentariolus*. I mean to say: without Tiedemann Giese you would not be here! I am quite certain you'll like the man, and not just because he is interested in my theory. He is a real human being, a true Christian.— By the way, he wants us to bring the manuscript and all our working papers along so that we would not have any excuse to leave early." Taking his seat again he started to talk about some details of the trip. We would, he said, take our time. Most of all, he wanted to show me Allenstein and its castle. "It's more or less on the way to Löbau, and I could imagine that you'd like to see where I spent a four-year term administering some large rural estates, the joint property of our Frauenburg chapter. The present administrator has been there for much longer, Achatius von der Trenck — not exactly the most urbane fellow you ever met, but well-meaning. He loves to entertain people with stories about his sister Mitzi — you'll hear some hilarious tales! Be prepared. Against the wish of her family Mitzi had married what I would call a wayward stargazer, an enthusiastic dilettante who imagined himself a genuine scientist. He never held a position and used his inheritance to satisfy his craving for beer and astronomical gadgets. His wife, Mitzi, ardently hated his hobby, because he spent more nights up on the roof of their house than in bed with her." Copernicus laughed and shook his head. "I am afraid there will be no new stories about him. On a star-lit night last February he suddenly died unexpectedly. Completely drunk he tried to make an observation, slid off the roof and was dead on the spot. Achatius jokingly claims that this was no accident, but that the Lord finally responded to Mitzi's prayers. In any case, at Allenstein Achatius will be our host for a few nights before we move south; by the end of the week we should be reaching the bishop's residence." Copernicus looked at me with a big smile and added that it still seemed unreal to him that Tiedemann Giese should be a bishop, whereas two years ago he was hardly more than a member of the Frauenburg Chapter, a friend and good neighbor. "But he certainly deserves this distinction, and I am very glad for him."

***
Diary Entry:
Glanden, July 28, 1539

On horseback all day. The heat was exhausting. Glanden, the hamlet we reached before dark, is the pits. Shacks in the middle of nowhere. Smelly dogs, muddy children, stooped peasants. What a life! What a place to spend the night!

My belief that we have a choice in life is being challenged by the hopelessness of this place. Could one really "transcend" one's lot, if born in such a place? How would one go about it? If I had been born and raised here, how would I get to Wittenberg? How would

I find my way to Nuremberg . . . and to my work in the tower by the cathedral? —- In spite of the smelly straw mattress I slept through the night without waking up once. But why this dream? Why did I have to dream about Virginia again? And why in this place? — My mother's letter? Marriage out of the question. Mother needs a reply. —- Watched Sebastian wash himself.

The Stranger behind the Copernican Revolution

Northern Poland 1539

50 Miles

From 1250 onward, Polish princes and kings invited German and Dutch peasants and craftsmen to settle in northern Poland. As they were permitted to retain their language and culture, new settlements were given German names. Polish names for these places were mostly coined with an eye on the German designation. The reverse was the case with existing Polish place names like Gdansk (899 A.D.) which became Danzig in German.

Allenstein — Olsztyn || Elbing — Elblag || Frauenburg — Frombork || Heilsberg — Litzbark || Löbau — Lubawa || Thorn — Torun.

II

TOASTING

THE

ASTRONOMERS

Allenstein Castle

The Stranger behind the Copernican Revolution

***

To his friend and colleague Paul Eber
in Wittenberg
Allenstein Castle, August 2, 1539

In the company of my Mentor I am presently on the way to visit one of his old buddies, a bishop, anxious to meet a mathematical genius from Wittenberg! — Just kidding, of course, but let me tell you, Paul, how lucky I was to receive your letter just two days before our departure. As you might guess, I was thrilled to find the excerpts from Lemnius' new satire on Luther. As you may guess, I have been waiting for them rather impatiently. Reading the verses I had the time of my life! His wit! His images! And what a revenge on our revered Reformer! (God bless him!) I loved in particular the bedroom scene where Katharina Luther, frustrated by Dr. Martinus' dysfunction, decides to get a student 'stud' in town. In this passage Lemnius has outdone himself! Poor Dr. Martinus! If he ever came to know that you, Paul, helped the author to escape from Wittenberg and are now spreading his indecencies, you would need the combined protection of Heaven and Hell to save your skin. The title *Monachopornomachia* certainly fits. Scultetus, to whom I read your selection, was laughing his head off. Under any circumstances, send me a copy of the book as soon as it becomes available.— Who cares whether all this is fiction or fact as long as it is outrageously funny and grossly indecent. At a propitious moment I might even show it to Copernicus.

Katharina Luther, née von Bora

A nun at age 16, a runaway nun at 24, who married a religious rebel and former monk two years later, she was an easy target for Simon Lemnius' scathing sexual and scatological satire that appealed to his sixteenth century audience.

To my surprise I learned through the grapevine that my revered Dr. N. has produced a piece of erotica himself, a translation of a "Greek author of dubious taste and doubtful reputation", whose name I had not heard before —- Theophylaktos of Simokattes. As Scultetus puts it, "Coming fresh out of Italy, Dr. N. blithely touted this piece of porn as a 'southern delicacy' and finally, in

1509, had it printed with a dedication to dear Uncle Lucas, Bishop of Varmia, who apparently lapped it up in one session". I haven't seen a copy of the book, and I suspect that Scultetus exaggerates its lubricity. In all likelihood, it is no match to Lemnius' salacious filth. When I read how you engineered the extension for Heinrich Zell's stay with me, I could have jumped up from my desk and sung your praises to the waters of the Lagoon. Your diplomatic skill is amazing, and to me it is a boon indeed. I can assure you that Henricus will continue his work on our map while I'll try to finish my *Narratio,* the summary of Copernicus' book. By November I hope to have found a printer in Danzig where Henricus, eager to shack up for the winter months, will take care of the proofs. The study and revision of my Mentor's manuscript will probably keep me busy for another year, possibly two. There are just too many details to be checked and confirmed, but don't feel sorry for me; this work is unbelievably exciting. Your question as to when I'll be in Königsberg I cannot answer at this time. But I will definitely want to see Duke Albrecht and that, of course, with the help of our old friend Spindler. — Please write again and don't forget to remember me to Lemnius when the occasion arises.
                      Your friend Joachim.

Martin Luther

The Stranger behind the Copernican Revolution 51

***
To Johannes Schöner,
astronomer and geographer in Nuremberg
Allenstein, August 2, 1539

Sheltered from the sun, I am writing in one of the corners of a courtyard in the Allenstein castle. Our host, with the uncommon name of Achatius von der Trenck, is an obese and jovial fellow in his forties with rather unrefined manners, awful eating habits, and all the conventional doubts about a moving earth. At the same time, I must confess that he is quite likable, but his habit of calling my Mentor "Nic" gives me a jolt in the stomach every time I hear it. I know, this shouldn't bother me, especially if I tell myself that the two have known each other for twenty years, but —- who wants to be rational *all* the time?

We arrived here three days ago, and Dr. N. has been reveling in his past. This is the place where the canons of Frauenburg take turns residing as overlords of the estates owned by the church in southern Varmia, and during Copernicus' tenure at Allenstein, Luther's name became a household word here. It was in this place, he told me, that in 1518 he received a copy of Dr. Martinus' famous *Theses Against the Sale of Indulgences,* and like other Frauenburg canons at the time, he felt for the man! — With your knowledge of history you may know that the Allenstein castle was built by the Teutonic Knights a few hundred years ago. It is not their biggest nor their most beautiful fortress, but within its walls these canons now live like princes. I am hardly exaggerating when I say that a canonry here would be a splendid thing for your friend Rheticus. With servants to take care of almost everything, I could indulge in my research undisturbed. I would even be willing to put up with an occasional assault by the Teutonic Knights, whose power, by the way, has been declining ever since they lost Varmia and West Prussia to Poland in the last century. In 1520, when they besieged Allenstein, my Mentor had to mobilize his skeleton crew for defense, but in his modesty he shrugs off any credit for what his bishop later extolled as an "act of heroism". If the Knights had seriously tried an attack on the castle, he argues, they would have taken it in a wink.

As you may guess from my sitting outdoors, our weather is better than one would expect here. But I am warned that this "touch of paradise" won't last. According to Achatius von der Trenck winters are rather "arctic" in this region and wolves abound. — This morning Achatius left to settle an old dispute between feuding peasants, and my Mentor opted to sleep in, a most prudent decision in view of all his tippling last night. Ever since the beginning of our journey he has been in a marvelous mood. The prospect of seeing his old friend Tiedemann Giese is visibly revitalizing him. I have never seen Copernicus as communicative as this, and last night I had the surprise of my life when, in the presence of our host and two guests, he suddenly started to reveal more about himself and his work than I had been able to extract from him since my arrival in Varmia. It seems this man will, in some way, remain an enigma to me.

The scene of last night's conversation is still fresh in my mind, and I must put it in writing before the colors fade. It might well become one of the most interesting parts of my *Vita Copernici*. — Meanwhile — with the manuscript of *De Revolutionibus* in my baggage, and in the company of its author — I am, and hope to remain, the happiest creature in Varmia.

J.

***

To be included in *Vita Cop.*:
Allenstein Castle, August 3, 1539.

For the evening meal Achatius von der Trenck had invited two of his friends from town, a lawyer and a priest, both men in their forties. After a light desert our host suggested to take advantage of the warm weather and sit outside. On our way to the courtyard he took the priest by the arm and, somewhat inebriated by now, asked him whether it wasn't odd for a professor to come all the way from Wittenberg to be assured that our earth was not firmly anchored at the center of the universe. And with a peculiar drawl in his speech he added, "And this afternoon our Wittenberg guest gave me a lively demonstration of how the earth has traded places with the sun and is now merrily orbiting that shiny ball of fire —- turning, at the same time, like a piece of meat on a roasting spit." He burst out laughing and pounded the priest's shoulder as they walked out into the courtyard.

"I wouldn't mind a moving earth" said the priest, "if the Bible were more explicit on this point. You remember, of course, that Joshua commanded the *sun* to stand still — not the earth! If the Scriptures are the revealed Truth of God — and I do not doubt that they are — why did the Lord, or the Holy Ghost or whoever . . . not tell us?"

"You forget", said the lawyer, "that the Bible was written or dictated to the prophets at a time when the Children of Israel were simple herdsmen with little or no education."

"Sure enough", said Copernicus, "but how much further are we today in spite of all our learning and erudition? What do you think people would say, if the Holy Ghost descended from Heaven right now to announce loud and clear that the sun was standing still and the earth was revolving around it? Can't you hear them snicker and say, 'Didn't I see the sun rise this morning making his way across the sky . . . just as the book of Ecclesiastics tells us?"

By the time we had reached a spacious arbor in which servants had placed additional chairs and readied two lanterns to be lit at the onset of dusk, I was ready to comment on the complete compatibility of Dr. N.'s theory with the Scriptures, summarizing the main arguments of the essay I had sent to Melanchthon, but Achatius burst out again, "Joshua, of course. As you know, Book X, verse twelve! And Ecclesiastics I:4, 'The sun arises and the sun goeth down, but the earth abideth', or something like that. If you have any other quotes, let me have them now." While everyone smiled,

The Stranger behind the Copernican Revolution 55

Achatius reached for the wine that a servant had placed in front of him. He suddenly looked quite serious, but instead of bragging with other biblical quotations, he turned to Copernicus, saying, "True as your theory may be, my dear Nic, the sad fact remains that you have no scientific *proof* of a moving earth — or did I misunderstand you?"

There was a pause, during which all attention was fixed on Copernicus. The faint ironic smile on my Mentor's face disappeared when he said, "I have no valid proof. But I wouldn't call it a *sad* fact. It simply means that science today has not advanced far enough to let us produce unequivocal evidence of the earth's motion. However, this uncertainty is a tremendous incentive for people like my friend Joachim or myself; and it certainly does not mean that the earth *isn't* moving. In time, of course, sustained effort will bring us closer and closer to the truth, but at this point there is very little any astronomer can *prove*. To be honest, there is only one thing beyond any doubt, namely that the angular distance between any celestial body and the horizon constantly changes. Everything else is subject to debate or interpretation of data."

"But what does this change of angular distance tell us about our position in the universe?" the priest asked. "I don't really understand." — Copernicus calmly replied, "It means that either the heavens are in motion — or the horizon — or both. We don't know, as yet. And as nobody has ever seen the earth move, people simply assume it to be fixed."

"But our senses", the priest said, "were most certainly given to us by the Lord to understand this world. As a gift of God they ought to be trusted."

"Naked eye observation can certainly trick us", Copernicus replied. "If you are sitting on a boat close to a pier in quiet waters and imperceptibly the boat starts moving, you can easily gain the impression that it is the shore which moves."

"True enough", interjected Achatius. "But you will not be fooled for long!"

"No doubt about that!" replied the lawyer. "And you know, of course, why!" said the priest, "The proximity to the shore makes it easy for you to check what's going on — and you can always rock the boat!"

"But the sun is thousands and thousands of miles away", said Copernicus, "and to rock the earth you'll need a little bit of help." He smiled and gave me a wink.

"What astronomers seem to need", said the lawyer, "is some kind of device to sharpen their vision, something like those spectacles that people are using nowadays." He wanted to elaborate, but our host interrupted him again.

"What I would like to know", said Achatius looking straight up into the sky, "what I would really like to know is this: how does an intelligent human being — a learned man like our friend Dr. Nicolaus — come to propound a theory that seems to defy all logic?" Without looking at anyone, still gazing into the sky, he went on, "I am sure there is some method in Nic's madness. And before I say more, let me tell you that I have

The Stranger behind the Copernican Revolution

always admired him for seeing this thing through . . . with all these calculations in trigonometry of which I understand so little. — But what I would like to hear is something very specific." He raised himself up, and now looking at Copernicus, said, "Could you, Nic, somehow pinpoint the beginning of your revolutionary idea? I ask myself: Was there an unexpected insight, a kind of revelation that made you suddenly realize that we live, or *might* live, in a sun-centered universe?"

Everyone's eyes were fixed on Copernicus who sat there motionless, looking at his goblet as if he had not heard the question. I expected him to be as evasive as ever on the issue and tried to anticipate his reply, when the lawyer finally broke the silence. "From my own experience", he said, "I know that solutions to difficult legal problems sometimes seem to strike me out of the blue, provided, of course, that in the preceding days or weeks I have struggled with the problem."

"I know what you mean", Copernicus said calmly, "because I have experienced that a number of times — but not with the idea of the heliocentric universe. However, I could imagine that Aristarchus, the Greek astronomer, mentioned by Cicero, was struck with the sudden insight of a heliocentric universe in which the earth moved like a planet — but in my case it was a combination of many things which were in the air — in the sciences, in society, in our Catholic Church. If I look at my work very soberly I would say that all of the thinking and all of the calculations and observations could have been done a hundred or three hundred years ago. But if I had lived at that time and had been blessed with a Frauenburg canonry, I would — with utmost certainty — not have written that book. Even with the same interests, the same instruments, and the same amount of leisure I would not have found this topic worth investigating. No, looking back on my own development, I would say that the most decisive stimulus for my venture came from a marvelous teacher of mine, that Dominico di Novara at Bologna. As a hotheaded dynamic Italian, he loved to shock us with irreverent remarks about Aristotle and Ptolemy, forcing us to think critically. At times he would show us why he considered Ptolemy the most creative mind in the history of astronomy, and then he would turn around to demonstrate the weaknesses of the Ptolemaic system. I remember one lecture in particular in which he elaborated on Ptolemy's singular ability to visualize spatial relationships, a gift that enabled him to explain the strange behavior of most planets: why these wandering bodies would suddenly stop, regress, swing into a loop and finally continue in their orbits. Using all kinds of drawings and self-constructed machines, Dr. Dominico would dramatize in a rather funny way Ptolemy's use of deferents and his practice of heaping epicycles on epicycles. He would joke and laugh at the turning and grinding of his celestial wheels and, in the end, derive great pleasure from seeing all of us puzzled and confused. He would finally pause, look around and say, 'In Ptolemy you see the workings of a superior intelligence. That's what I call a genius, that's what I call a creative

The Stranger behind the Copernican Revolution

Vaulted Room Allenstein Castle

thinker. Ptolemy was by far the most creative mind I know. But what you have seen today is by no means the culmination of his achievement. Next time I'll surprise you with a demonstration that will leave you speechless. You'll see how Ptolemy solved the age-old problems surrounding the sun and the inferior planets by tying together the deferents of Venus, Mercury and the Sun so that the center of their epicycles always lie on a straight line between the earth and the sun. This device alone would have brought immortal fame to Ptolemy. It was the triumph of the human mind over the intricacies of the heavens. Finally everything in the universe made sense! There was nothing in the sky that could not be explained with Ptolemy's teachings. Man had conquered the heavens ... conquered the universe.' — With all of us sitting in awe and amazement, Dr. Dominico gave the impression of being in a state of rapture over the flawlessness of Ptolemy's system. But then an ironic smile appeared on his face, and almost whispering he would say, 'This state of perfection lasted no longer than a few centuries. Then, the flaws of the system became apparent, and — lo' and behold — the patchwork began: to all the existing deferents and epicycles new ones had to be added. Instead of becoming simpler, the system grew more and more complex.' Novara's large brown eyes were now beaming with glee and he paused again to heighten the tension in the room. Finally he said, 'Now look at the mess we are in after more than a thousand years of astronomic meddling: we are still at a loss to explain the changing light intensity of Venus, we are still debating the orbital order of the lower planets, while in the meantime the seasons are getting more and more out of kilter. By now, spring comes ten days too early, and the holiest of holidays are falling on the wrong days. You are eating meat when you should be fasting, and you are fasting — what a shame! — when you could fill your belly with all the goodies that God provides us with. You do not need me to tell you that the system is flawed — in spite of the brilliant ideas that went into its construction. The time is ripe for revision — but who will be able to tackle the problem?' At this Dr. Dominico would pause and let his eyes wander across the room, before asking, 'Is there anybody here who still believes in the infallibility of the Ptolemaic system?' Of course, none of us would admit that our trust in Ptolemy and Aristotle had not really suffered under the impact of Dr. Dominico's rhetoric. He paused, once more, waiting for a reaction. 'I am pleased,' he finally said, 'to have before me a group of reformed astronomers! The only disturbing thing is that I have given this same little sermon for years without ever seeing a second Ptolemy emerge from my classes. But today I have a new challenge — and a serious one! I invite all of you to witness an extraordinary event in the heavens, an occultation of the star Aldebaran by the moon. This is a rare opportunity for us to test the validity of Ptolemy's observations and theory of the moon. Join me — I am serious about this — join me tomorrow night March 9, at eleven o'clock for a very unusual session, and let us all pray for a clear sky, Amen.'

Copernicus paused and looked around, apparently waiting for a question. "That was in 1497", he finally said. "Of all the students in that class, only two appeared for this observation, a Frenchman by the name of Émile and myself. Novara was very excited that night, but he worked with utmost care and circumspection, measuring and evaluating all data, checking every step several times to make sure that no mistake would creep in. Then, with the final results in hand — and once more checking every step — we came to the conclusion that Ptolemy must have been wrong in some of his basic assumptions about the moon. This did not disprove anything regarding the system as a whole, but to experience in this way the fallibility of that great man was somehow shocking and, at the same time, exhilarating. However, what impressed me more than anything else during that night was our final conversation with Novara. We were sitting in a corner of the observatory at a small table with an armillary sphere standing in the center. It was almost morning; I could see that Dr. Dominico was completely exhausted, but there was still something on his mind that prevented him from dismissing us. 'Do you think,' he finally said, 'that Ptolemy's choice was a good one, when he decided to build his extraordinary universe on Aristotle's model?' While the Frenchman and I exchanged questioning glances, Novara looked at the armillary sphere in front of him and started playing with the rings as if he were all by himself. With none of us willing to speak, he finally said with a sudden turn of his head, looking me in the eyes, 'I would be interested in getting an answer to my question.'

'The result of our session', I said with much hesitation, 'seems to corroborate what you have said a number of times: that Ptolemy's choice of an earth-centered universe was *not* a good choice.'

'I don't think I ever said such a thing.' Novara replied with a smile, 'even though you might have had that impression. — But let's hear Émile's answer to this.'

The Stranger behind the Copernican Revolution

Shrugging his shoulders and pointing at the spheres in front of us, Émile replied, 'This model looks like a good choice to me.'

'You said it!' Novara retorted visibly pleased. 'It *was* a good choice! The system worked so well that we are still using it!' And taking a deep breath he added, 'Considering the state of astronomy in the second century A.D., Ptolemy probably made the best of all possible choices. But that does not mean that his model of the universe corresponds to the actual structure of the heavens. Things may look quite different up there from what he imagined. The flaws that we see today are, of course, the benefit of hindsight. No person in Ptolemy's time could ever have foreseen the disarray we are experiencing in the system today. In short, the best choice — in Ptolemy's case — was the wrong choice'." Copernicus paused and everyone seemed to ponder over the last sentence.

"What an amazing insight!" the lawyer finally commented. Copernicus nodded. "Strangely enough, Novara never came to know the impact of this remark on my life. To him it was not more than a casual comment, because months later, when I went to see him about it, I did not get much help. Nevertheless, his apodictic statement became an obsession with me. I was haunted by it and haunted by the question it raised: if the best choice had been the wrong choice, which choice would have been the right one? Finding no fellow-students seriously interested in the problem, I was thrown back on my own cogitations. I often imagined how God, the master builder of the universe — before the first day of creation — might have envisioned the world He was about to create. And many times I asked myself: what would He have preferred to see at the center of His prospective universe? — Pondering that question, it seemed to me that either the earth or the sun might easily have been favored by God. The earth, because it was to become the home of Adam and Eve and eventually of all humankind, and as such it was destined to become the Lord's primary center of attention. On the other hand, I could also see that He might have preferred the sun in central position as it was to be the brightest and biggest object in the heavens with all the smaller ones revolving around it. In that way the sun, the source of life-giving light, would automatically become a symbol of the creator Himself. My strongest objection to placing the sun in that central position was, of course, my own upbringing, my traditional thinking and our perception that all heavenly bodies are revolving around our solidly anchored earth. But over the years I came to accept the fact that the constant change of angular distance between the horizon and the heavens was no proof of the earth's stationary position.

The only other person truly impressed with Novara's magic phrasing was my friend Tiedemann Giese, from whom, at that time, I received an enthusiastic letter about his classes in trigonometry and astronomy. As a response I sent him an account of our nocturnal session with Novara, and on a separate sheet I printed in huge decora-

The Stranger behind the Copernican Revolution

tive letters Dr. Dominico's paradoxical statement: *'The best choice — in Ptolemy's case — was the wrong choice.'* — Within two weeks I had an answer from Tiedemann, and from that day on we constantly exchanged letters. When I finally left Italy and saw Tiedemann again, we saw each other whenever he came to Heilsberg. He was and is an imaginative thinker who would often raise questions that I had never thought of, but he always insisted that he did not have the mathematical gift to do anything seriously in astronomy. Yet through him and his questioning I became, once again, so involved with the problem of a moving earth that I finally made a serious attempt at charting in my imagination — and on paper — the course of Jupiter as it would appear to an observer on earth, imagining both Jupiter and the earth in orbit around the sun: Jupiter taking almost twelve years to complete the journey, the earth only one. The result of this thought experiment and the ensuing calculation was stunning.

I found that an observer on earth would see Jupiter stop in its forward movement, regress and go into that familiar loop *without* the help of Ptolemy's deferent and epicycles."

"You mean to say", shouted Achatius excitedly, "that Ptolemy's rotating wheels in the heavenly spheres are applesauce and rubbish?"

"I wouldn't put it that way", said Copernicus, "I would rather call them ingenious devices to explain the irregular movements of the planets."

Still puzzled, Achatius looked at me, shaking his head. "I don't understand."

"You need these devices", I tried to explain, "if you assume that Jupiter is orbiting a stationary earth. If, on the other hand, you assume that the earth and the planets are orbiting the sun, things look much neater. In that case the retrograde motions — these strange loops that have baffled observers throughout the ages — turn out to be the result of the earth's journey around the sun. This is in fact the most compelling reason for our assumption that we live in a sun-centered universe."

"Let me repeat", said the lawyer with an engaging smile. "The loops are simply a result of the earth's journey around the sun?" When I nodded, he became ecstatic.

"That is amazing, that's unbelievable! Things are really less complicated than I had thought, but to come up with this idea and be able to demonstrate it —- what a tremendous feat!" He stood up, took a step toward Copernicus, lifted his glass and said, "You must have felt an immense joy and elation when you came to see that these loops were simply caused by the motion of our earth. I must salute you!"

Copernicus smiled. "I was thrilled, I was overwhelmed when I saw things falling into place. It happened in the summer of 1506 during a leave which Bishop Lucas had granted me for pursuing my interests in astronomy. To avoid the distractions of the court at Heilsberg, I went to Frauenburg to stay with Tiedemann Giese. He was perhaps even more anxious than I to see a solution to the problems that we had discussed over the years. Although I worked very intensely during the first week or so, I seemed to make no progress at all.

Then I decided to explore the unlikely possibility of a the earth moving around the sun, as I just mentioned a minute ago. I was still very skeptical of the outcome, and for that reason I did not share my thoughts with Tiedemann until I could demonstrate to him that an observer on earth would see Jupiter in retrograde motion while in fact the planet was moving undisturbed in its circular orbit around the sun. At that point I was ecstatic. I grabbed my papers, rushed downstairs and burst into Tiedemann's study with my incredible news. It was perhaps the happiest moment in my life."

There was a long pause which the lawyer and the priest filled with slow repeated nods. Suddenly Achatius broke the silence with an embarrassing snicker, but sensing his gaffe he apologized and explained that he had pictured his late brother-in-law, Hubertus, being a guest at this party.

"That would have been a melodrama", said the priest, dispelling a certain uneasiness.

"I agree", said the lawyer, "we would have been well entertained. Hubertus was by far the strangest man I ever knew. One might justly say that he was everything that Dr. Nicolaus is not: a dabbler in astronomy with an inflated ego and a burning ambition to be famous. May God have mercy on him."

"You forgot to mention his bad temper", interjected Achatius rising to his feet. More and more inebriated by now, he spoke as if he were addressing a large audience. "As some of you know", he said, "nobody suffered more than my sister Mitzi. Hubertus blamed her for just about everything that went wrong in his life, even for his failure as a scientist. Can you imagine that a fellow would travel all the way to Rome on three or four occasions with some silly proposals to change the calendar?"

With an ingratiating smile the lawyer turned to me saying, "You may certainly get the impression that our host is laying it on; I assure you, he is not. Hubertus was my closest neighbor for twenty years, and I saw a lot of him. If he had been with us tonight, he might well have had one of his tantrums."

"No doubt about that!" cried Achatius, "He would immediately claim priority over Nic's discovery, and would beat his chest for not having published his findings."

"On such occasion", the lawyer added, "Hubertus could lash into a fury, smash mugs and furniture, and be on a beer binge for the next two days."

The priest nodded and smiled. "That's the way it used to be, but the Lord called him home." There was a sudden silence, and everyone looked down, as if that was the appropriate pose to ponder the fate of Hubertus. With a trace of irony in his voice, the priest finally said, "God bless his soul." We all looked at each other, and with a sigh of relief lifted our glasses to drink in silence.

Two servants busied themselves refilling the goblets, when the lawyer rose to his feet to suggest another toast to Dr. Nicolaus and his discovery.

"Please don't." said Copernicus firmly. "I am all in favor of toasting, but not to my work or my person. I just don't like it. But I wish we could salute all the great men who have made my work possible, dozens of them from Anaximander and Democritus to my young friend Rheticus. You can be sure that without his help I would not be preparing my manuscript for the printer."

Achatius was elated. He immediately asked for two musicians to make their appearance and gave orders to bring out a special case of French wine for which he claimed to have paid an exorbitant price. "Regardless of the earth's position", he shouted, "all these illustrious observers of the skies ought to be honored!" And before anybody could say anything, Achatius suggested to toast them in chronological order, starting at the dawn of civilization. "I am sure", he said, "there are some Assyrian stargazers whose names you remember". When none of us seemed to remember, Achatius proposed to drink to them anonymously. "After all, they gave us the Zodiac, and probably much more!"

While the lamps were lit, the musicians began playing softly in the background. Achatius lifted his goblet to the stars, saluting the unnamed Assyrians. I noticed that my Mentor skirted this absurdity by rearranging the cushion supporting his back. The priest visibly animated by Achatius' performance, turned to the lawyer and asked, "Where are we going from here?"

The lawyer's eyes lit up and an impish smile appeared on his face. "To my own surprise", he said, "the names of some Babylonian astronomers have come to my mind, don't ask me how!" Achatius burst out laughing and nobody doubted that the lawyer was pulling our leg. "You name them, and we'll toast them",

Achatius responded. "And after that we'll do the Egyptians, the Greeks and all the others. I can see, we're going to have a ball."

From this point on, my Mentor drank hardly anything. When the party finally ended, he had no difficulty getting up from his chair. I was amazed to see him walk away poised and steady, needing no help from his valet or the torch bearer.

***

To be included in *Vita Copernici*.:
Allenstein Castle, August 6, 1539.

I was packing my bags, getting ready for tomorrow's departure for Löbau Castle. Enter Dr. N. to tell me that we would be eating dinner at the house of an old acquaintance, a former Königsberg doctor and staunch Catholic. "The man left his native city for religious reasons." Dr. N. explained. "When Duke Albrecht sided with Luther and the Königsberg bishop followed Luther's example and got married in 1525, he moved to Allenstein. I am sure you'll enjoy his company." He paused and paced the floor for a while. "But I also came for another reason", he added in a businesslike manner. "Before we get to Löbau, you ought to know a little more about Tiedemann Giese. I've mentioned, of course, that he comes from Danzig, but I have to add that the Gieses are a very influential family . . . old money and sound politics. His father was mayor of Danzig and rel-

The Stranger behind the Copernican Revolution 71

atives on both sides still hold important positions. By the way, our last bishop of Varmia, Maurice Faber, who died less than three years ago, was his uncle and had, as things go in Prussia, everything prepared for his nephew to succeed him. Unfortunately, the King wanted Dantiscus, and so Tiedemann was given the diocese of Culm. To be honest, Culm was a poor second, a kind of consolation prize, with much less political clout and less income. I thought that knowing this might be helpful in avoiding embarrassing questions. As you might guess, jealousy in the ranks of the upper clergy is as rampant as in academe." He looked at me smiling somewhat self-consciously. While he continued talking, he stepped up to the window that faced the park. "I have known Tiedemann from early on, and as canons at Frauenburg we have been together since the death of my uncle. If I ever had a friend, it was this man; and if my book amounts to anything, he certainly has a share in it. Through his prodding and persuasion I kept on working, and many a time he was an inspiration to me." He suddenly paused, and when he continued a slight tremor shook his voice. "At times I feel guilty that I did not encourage him enough with *his* book *On the Reign of Christ*. Unfortunately, theological speculations are not my forte. In more than one way they are beyond my grasp. But a number of prominent theologians have urged him to publish it, and even Melanchthon has seen and praised the manuscript. It probably is an excellent book, and in no way controversial."

"What then holds him back? Do you understand?" I asked.

"Dr. Tiedemann is such a perfectionist that he starts revising as soon as he takes another look at it." Copernicus chuckled. "Scultetus used to make fun of both of us, pointing out that I have been sitting on my script just about as long as Tiedemann has been 'squatting' on his. He used to claim that we were in love with our books and for that reason would not part with them."

"But seeing your book on the way to the printer", I said, "may be enough to encourage the bishop."

"I doubt it. He is now restructuring large sections, and who knows what will be next!"

"Is it true", I asked, "that he always keeps the two copies of his manuscript in different places? I heard that from Scultetus."

"All his friends tease him about that" said Copernicus, "and like Petrarch, who kept a cat to protect his manuscripts from mice, Tiedemann always had a feline companion."

I burst out laughing. "You might not believe this", I said, "but as a boy I actually saw the skeleton of Petrarch's famous cat. I was ten years old when my parents took me along on a tour to southern France where my mother insisted on seeing Petrarch's house at Vaucluse. As a true-blue Italian, she was in love with Petrarch's poetry and knew his most famous sonnets by heart. So, when she entered his house — finding it intact some hundred and twenty years after the poet's death — she was ecstatic. But then there was this zany caretaker who loved to surprise and shock visitors. So, while I was playing with a giant abacus on one side of the room, I hear, all of a sudden, my mother scream in panic and disgust. Turning around, I see her covering her face while this crazy guide is pulling the skeleton of Petrarch's old cat out of a chest, lifting it high up to the ceiling."

Copernicus laughed, telling me that during his student days at Bologna he had heard a similar tale. By the way", he added, "I appreciate your mother's love of Petrarch; but you have to tell the cat story to Tiedemann Giese. For the last five years he has had two cats, and in a mock-ceremony we dubbed them 'Wardens of the Writ.' Yes, Bishop Giese has a sense of humor. You'll see that when you meet his feline dignitaries." With a mischievous smile he added, "Their names have a distinct Wittenberg flavor and may offend high-minded Lutherans." I looked at him questioning and puzzled.

"The tomcat", he said, "is called Martinus and the neutered female Katharina." Visibly pleased that I was chuckling he moved toward the door. "As for tonight", he said, "there is no need to dress up for the dinner. The doctor is rather informal." Before I had time to say anything, Dr. N. was out of the room.

III

EUREKA

Tiedemann Giese

The Stranger behind the Copernican Revolution

***
To his friend and colleague Paul Eber
in Wittenberg
Löbau Castle, August 11, 1539

This "castle" is more an oversized residence of a wealthy merchant than a bishop's palace. Its architectural design would, in your parlance, be a consummate flop. The town itself, though larger than Wittenberg, seems intellectually dead, except for its very bright bishop. Fortunately, I see a lot of him. The enclosed list of items reflects his interests and wishes: books and pamphlets which his Excellency is anxious to lay his hands on. Knowing your reliability and speed, I was brazen enough to boast that I could get these things much faster to him than the local dealer. Looking a little closer at that list you might cry "heresy and dissent!", as most of the items are Protestant publications. Yet, Tiedemann Giese is not going to jump the fence. As an incorrigible idealist, he is still hoping to unite the two faiths.

I am only half joking, when I say that this man would make a superb Wittenberg prelate: affable, urbane and tolerant, unlike Luther in every respect! Not that I have anything against our illustrious Dr. Martinus ... at least not as a biblical scholar, even though I never appreciated his uncouth manners and his four-letter words at the dinner table. By contrast, Dr. Tiedemann might be a little too genteel for your taste, but certainly not for mine. Even looking at him is a pleasure. As my mother would put it, "One doesn't even have to get close to him to scent the patrician bouquet." Needless to say, he is a wonderful host, and for that reason the first week here has been a kind of gala party. To my surprise I discovered that nobody on earth has more detailed knowledge and more reliable information about my Mentor than this man. He witnessed Copernicus' struggles and some of his breakthroughs and has seen the stalling and gradual progress on his work. He favors my idea of a biography and is talking a lot about the years they spent together at Frauenburg. That my Mentor is not in town is a fortuitous coincident which opened up a rich biographical source for me. Today Dr. N. left for a number of days visiting an old friend a few hours east from here. This means we can meet undisturbed any time and I can even take notes while the Bishop talks. Needless to say that this turn of events has lifted my spirits high above the drab design of the Löbau Castle, and with a second bottle of wine on my table I drink to your health and my good fortune. How I wish you could see me at this moment, elated and thrilled sitting high on cloud nine —- so high in fact that you could barely recognize your old friend Joachim.

***

Diary Entry:
Löbau Castle, August 12, 1539

Lunch with Dr. Tiedemann; more notes for *vita Cop.* (1526-1530). —- Message from Spindler: Duke Albrecht received my letter and wants to see me in Königsberg. Good news. Excellent news! Must write to Spindler that he stop at Frauenburg on his way to Danzig. Meeting Dr. N. is a must for him.

When am I going to write that letter to mother? No more procrastination! —- No new excuses! You have to be candid! Spell it out? But *how* . . . without devastating her? What a problem! O God!

***

Diary Entry:
Löbau Castle, August 13, 1539

Dream about S. — Worked on *Narratio* all morning. After our midday meal the bishop suggested to take advantage of the balmy weather and sit on the terrace overlooking the park. He had indicated earlier that he was in the mood to talk about his early years with my Mentor. —-

***
To be included in *Vita Cop.*:
Löbau Castle, August 13, 1539.

While a valet opened a bottle of wine, Dr. Tiedemann recalled one of his first trips to Thorn, where he and his mother were visiting relatives who happened to be friends of my Mentor's family. Most interesting to me: some unsettling facts about Dr. N.'s brother.

"When I first met Nicolaus", the bishop said, "I was only seven, and I hardly dared to talk to him. He was twice my age, and to me he seemed more like an adult than a boy; the same was true with his brother Andreas, who was even older. The boys had lost their father, and somehow I envied the kind of freedom they seemed to enjoy. To me their lives were strangely alluring. Then, I overheard my aunt saying that Andreas was "guileful and shifty", prone to getting into trouble with the law. At that time I began to fantasize about their ventures, although I had virtually nothing to go by." The bishop paused and looked curiously self-conscious. "When you asked me to help you with information about Dr. Nicolaus' life, I seriously pondered the question whether I ought to tell you anything about this brother. It would be easy for me not to mention him at all, just as Nicolaus stopped talking about him many years ago. But to do that in this context would amount to a misrepresentation and distortion of certain historical facts, because in my opinion Andreas' life is somehow linked to Nic's decision to finally go to the trouble of testing his theory by writing that remarkable book. There was sadness in Giese's voice when he continued. "Unfortunately my aunt had a point in calling Andreas shifty. But his uncle and guardian, Bishop Lucas, would stubbornly deny the darker side of his nephew. Proud of Andreas' intelligence and good looks, he persuaded him to choose the Church as a career, even though it was clear from the beginning that Andreas was unfit for the cloth. In his bullheaded fashion Bishop Lucas prevailed upon the Chapter at Frauenburg to accept Andreas as member while he was still studying in Italy, and where he was constantly living beyond his means. When he finally returned from the south he had contracted an incurable disease. In spite of that his lifestyle remained rather worldly and rumors about his transgressions flourished —- founded or unfounded. In any case, things took a turn for the worse when he was charged with the embezzlement of funds that had been earmarked for the construction of a church. The details are irrelevant, but when the matter became public, Nic was shocked and utterly distressed." Tiedemann sighed, lifted his glass as if he wanted to drink and slowly returned it to the table. "Opinions are split on the nature of Andreas' disease. According to Nicolaus, who has had some medical training and practical experience, he suffered from leprosy, while others think that he was afflicted with the new 'French disease' which the physician and poet Giromalo Fracastoro described in beautiful Latin verse, *Syphilis sive morbus Gallicus.* Its publication, eight or nine years ago, was a sensation in

Frauenburg. Everyone read it —- although that was long after Andreas' death. If you know Frauenburg, you will not be surprised that a number of fellow canons saw in his suffering a proof of God's revenge, and during the final stages of the disease they voted for his removal from the Cathedral and his ouster from the diocese, citing the danger of contagion. Bishop Lucas was dead when this happened, but for Nicolaus, who somehow loved his brother, it was very upsetting, very humiliating. But as I see it now, the fate of Andreas and the distressing events following the death of his Uncle Lucas had a rather constructive side. They impelled Nic to finally use his time to tackle the difficult task of writing that extraordinary book and let him forget the mental anguish and disappointments of those years. As we often say up here, *'Labor vincit omnia'*.

But let me get back to my youth. After that visit to Thorn, I did not see Nicolaus again until he was nineteen. He had just finished his first year of study at the university of Cracow and was visiting his relatives in Danzig who are close friends of my mother's family. That was late in the summer of 1492. I remember it so well because of the sensational news that Queen Isabella of Spain had commissioned three ships under the command of an Italian captain to find a way to India by sailing west across the Atlantic! Two Danzig printers had come out with illustrated broadsides on the subject, and everybody was speculating what would happen to the ships once they were hundreds or even thousands of miles out on the ocean. Many people were convinced that sooner or later they would fall off the earth. Our parish priest openly condemned the venture in his sermons and predicted that the ships would sail straight into the jaws of Hell. — I remember Nic joking about such nonsense; but he was skeptical for another reason. One of his Cracow professors had discussed such an undertaking, but had come to the conclusion that the technique of shipbuilding was not far enough advanced for seagoing vessels to withstand the onslaught of the elements so far out on the ocean."

Trying to clarify the difference in age between the bishop and my Mentor, I asked, "If Dr. Nicolaus was nineteen at that time, you must have been a boy of just twelve. Amazing, how much you remember."

He nodded and smiled. Self-irony was mingled with pride when he said, "I was a rather precocious child. My family started to show off with me before I was ten; by that time I had memorized long passages from Virgil and Horace and was able to carry on a fairly intelligent conversation in Latin. That impressed people. At that age I had not yet experienced the limitations of my gifts and was quite conceited. I was seeing myself as a kind of *wunderkind*. So, when clergymen, scholars and dignitaries came to the house, I was always looking forward to my stunts of showmanship. By the time of Nicolaus' visit, my tutor and I were starting to pack our bags, because in the fall of that year I was to enter the University of Leipzig."

"At the age of twelve?" I asked in disbelief.

"Yes, a little early, I admit. As Nic was a kind of late

bloomer, he was — as he told me later — tremendously impressed to see a kid of twelve so far advanced. We exchanged a number of letters over the years, and after a visit to Varmia — on his return to Bologna — I saw him again in Danzig. When he arrived, he was walking on air, raving about a number of books on math and astronomy. I have rarely seen him as bouncy and buoyant again. His mind just bubbled with astronomy and Euclidian geometry. Hearing him talk, I was baffled and bewildered because I knew almost nothing about these subjects, even though I had two academic degrees by that time, a bachelor's and a master's. Somehow I felt tremendously challenged; his scientific knowledge was humiliating to me. My pride was hurt, and I suddenly had the desire to prove to myself and the world that I could master these things easily enough. So, with the help of some tutors, I devoted myself to geometry and astronomy for almost two years. Needless to say, I started with great enthusiasm and I am still surprised that my eagerness to learn did not really suffer when I gradually came to see my limitations. The fact that I was doing quite well, did not delude me. For the first time in my life I had to struggle, and the further I advanced, the more I realized that I would never amount to much, if I had the ambition to become an astronomer. That was a rather sobering experience for an ex-*wunderkind,* but it certainly helped me to mature. In addition, it enabled me to appreciate in later years Nic's work and the significance of his discoveries."

Ferrara

As Copernicus had spent much of his time studying astronomy, mathematics and Greek, he finally went to Ferrara (1503) to receive the obligatory doctorate in canon law. The university of Ferrara had the reputation of being rather lenient toward its doctoral candidates.

As the bishop paused, I asked if he had some intimation of the budding genius in his friend. He thought about that for a while and slowly shook his head. "What struck me most during that visit in Danzig was his passionate involvement with math and his boundless enthusiasm for the sciences in general. He was, as I said, on his way to Bologna and could hardly wait to attend again Dominico di Novara's lectures on Ptolemy and the structure of the universe."

Somehow confused about Copernicus' studies in Italy, I asked whether his goal had not been the study of canon law.

"That was his uncle's wish, and design", said the bishop with an air of barely disguised disdain. "Bishop Lucas had high hopes of launching his nephew on a glamorous career in the Church, envisioning him as his successor and whetting Nic's appetite for a cardinal's hat. That was, as I see it, a gross misjudgment of character, because nothing was further from Nic's mind than playing an important role in politics."

"But what about the time he spent at his uncle's court in Heilsberg?"

"During those years", the Bishop said, "he was indeed in a powerful position, but that doesn't contradict what I just said. From 1504 onward, Nic and his uncle entered an exceptional symbiotic relationship. Lucas Watzenrode was a pigheaded autocrat, but he had an infinite trust in his nephew's good judgment. Whenever he had maneuvered himself into an awkward political position with the King or the Prussian diet, Nic would find a way out. And in domestic matters he was a master at calming the furor

The Stranger behind the Copernican Revolution

and inner turmoil of the old man. Through his rational approach to problems he managed to cajole Bishop Lucas into making amazing compromises."

"That is fascinating", I exclaimed. "I can see the detached scientist pulling the ropes and invisibly influencing Varmian, Prussian and Polish politics."

The bishop nodded. "That's the way it was. But after his return from Italy and his rise to power he had inadvertently made enemies, people whose career prospects faded, when Nic appeared on the scene. They had no means to fight the bishop's decision to advance his nephew who, as they saw it, shrewdly manipulated the bishop. Their greatest fear was that Nic would manage to hold on to his power after the bishop's death."

Lucas Watzenrode

"Did Bishop Lucas ever realize", I asked, "that his nephew was a scientist at heart, a scientific genius with extraordinary gifts?"

"I don't think so. Hardly anybody in Varmia did or does — even today. I would say that Bishop Lucas was quite impressed with Nic's scientific knowledge, but he wanted his nephew to become a refined version of himself, an ambitious politician with an eye on the Curia Romana."

Tiedemann Giese paused, apparently waiting for a reaction.

"I can understand Bishop Lucas' disappointment", I said. "But I have asked myself many times why a man like Dr. Nic has not made any attempt to move to one of the great intellectual centers where students would have profited from his erudition and the world would have celebrated his achievements. For a man like me, who wants to see the results of his research in print as fast as possible and who wants to be remembered by posterity, this is hard to grasp."

"I know what you mean", replied the Bishop, "But it seems to me that our friend Copernicus is an exception among scientists and men of great genius. — Knowing him well and looking back on his life, I can say with conviction that all his needs were met when his uncle set him up with a canonry at Frauenburg at the age of twenty-four. The position meant financial security for life, combined with an inordinate measure of personal freedom to pursue his scientific interests."

"Are you saying", I asked, "that he *preferred* the

unassuming rank of a canon in a remote little place like Frauenburg to a powerful position at his uncle's court in Heilsberg?"

"Perhaps not at that time", the bishop replied, "because of the novelty of his position. Those eight years were filled with extraordinary happenings and exhilarating events. Lucas Watzenrode was at the height of his career as a statesman. He was by then one of the most influential, most active and most controversial figures in Polish politics, and Nic certainly enjoyed his roles as adviser, personal physician and private secretary to this difficult man who would listen to him in moments of crises. Besides that, his uncle was extremely generous with him, and I might add that he loved the amenities and luxuries of the court. But my point is that he did not need the court at Heilsberg for his inner happiness nor for physical comfort, because at heart he always was what he still is: a rather withdrawn scientist with amazingly little need for public recognition." The bishop paused as if he wanted to give me time to ponder the matter.

"The year 1512", he finally said, "was a tremendous shock to Copernicus. But at the same time it was a wake-up call for the scientist. On the day Bishop Lucas died, Nic lost virtually everything that had shaped his life since his return from Italy. He not only lost his protector and second father, he lost his political clout and his social prestige. It was the day his secret enemies had been waiting for. The stage had been set to oust him as soon as Bishop Lucas closed his eyes. They had expected resistance, and plans were in place to deal with Nic. But he surprised them when he quietly left the castle and settled in Frauenburg, where —- twenty-seven years later —- you were still able to find him as a canon." With an ironic smile Tiedemann Giese added, "This coup d'état was probably the smoothest transition the diocese ever experienced in its long history, because those snakes in the grass had thought of everything. All the steps toward the nomination of the next bishop had secretly been taken, and an interim administrator stepped in immediately. Three days after they buried Bishop Lucas, the election of his successor was held, and all that remained, was the King's confirmation."

I was stunned. All this was completely new to me. I looked at the bishop and shook my head.

"It was traumatic for Nic", he continued, "but in addition, his brother became more and more an embarrassment.

"Under these conditions", I said, "the temptation to leave Frauenburg must have been great."

Giese smiled and shook his head. "I know him well enough to deny that. In the final analysis, Frauenburg meant contentment to him, at times even happiness in spite of it all. In Frauenburg he was free to follow his many interests and devote himself to whatever he chose. His *magnum opus* is evidence that he did not need the scientific community as an incentive or sounding board. A few friends who would listen to him and with whom he could argue about certain

The Stranger behind the Copernican Revolution

Castle at Heilsberg
From this castle Copernicus' uncle, Lucas Watzenrode, ruled Varmia from 1489 to 1512. Erected by the Teutonic Knights in 1241 the castle served as episcopal residence from 1315 to 1795. — After Copernicus' return from his Italian studies (1504), he moved to Heilsberg to become the Prince-Bishop's most trusted advisor, secretary and physician.

aspects of his problems was all he wanted. Apart from myself, there were two other canons who admired him, Scultetus and Dohner. There was no yearning for fame, no strong desire to bring his findings to the attention of others in the field, but there was, of course, the fear of criticism and ridicule, especially from people without any mathematical knowledge. And one has to admit that at first sight the theory of a moving earth sounds absurd to most people. You cannot imagine the effort it took to persuade him to send a copy of his *Commentariolus* to a few astronomers and mathematicians, and I shudder to think what might have happened to the manuscript of *De Revolutionibus*, if you had not arrived on the scene."

At this moment a scribe delivered a letter which Dr. Tiedemann opened and read on the spot. I recognized the seal of Dantiscus, and waited impatiently for a comment. The increasing tension in Giese's posture was obvious, his face flushed. "It's unbelievable!" he cried excitedly. "This is Dantiscus at his worst: In his crusade against wantonness and moral depravity he is losing all sense of propriety, making mountains out of molehills. His feud with Scultetus has been going on for more than a year now, and as far as I can see, it is much more a personality clash than anything else. Under the guise of moral fervor, he is now initiating Scultetus' expulsion from Varmia and is urging me to persuade Dr. Nic to endorse his removal from the Chapter. This is preposterous!" He stood up and nervously paced the terrace.

I told him that I had spent the evening before our trip with Dr. Scultetus, who had talked at length about his battle with the bishop. "He is seriously thinking of taking his case to Rome", I said, "but I have no idea what his chances of winning are."

"He still has some powerful friends in Rome", Giese replied, "and he may in the end prevail. But what upsets me is the brazenness of Dantiscus, who expects support from an individual whom he has just stopped harassing.

The Stranger behind the Copernican Revolution

You've probably heard that he pressured Dr. Nicolaus to dismiss his very able housekeeper. Wiping the sweat off his forehead he turned to me with a smile and apologized for his emotional outburst. Realizing that he was not in a state of mind to continue with his recollections, I thanked him for his trust and returned to my room, deeply concerned about the fate of my drinking companion in Frauenburg.

***
Diary Entry:
Löbau Castle, August 14, 1539

A letter from Dr. N. that he is recovering from a cold, which might delay his return. Looking forward to seeing him . . . and Sebastian. — Dr. Tiedemann out of town. — Discarded another draft of the letter to Mother. Too long — much too long and evasive! Try again. — Good progress on *Narratio Prima*.

***
To his Mother,
Thomasina de Porris in Bregenz
Löbau Castle, August 16, 1539

Almost three weeks ago my Mentor and I left Frauenburg to visit one of his oldest friends, a former colleague, who became Bishop of Culm a few years ago. If I were to describe him to you, I would never do justice to this impressive pontiff of Roman vintage. Let it suffice that I am happy and that you'll be getting the scoop on him in my next letter. As we'll be spending another five or six weeks with our gracious host, please write to Löbau Castle.

What is on my mind today and what has bothered me for some time will not be welcome news — rather sad tidings for my mother, the matchmaker. Yet, for my own sake and my relationship to you, I have to spell it out. When I saw you in Bregenz last year, I was convinced that you had understood my position on the delicate question concerning Virginia, and I firmly believed that you had put the matter to rest — excised, so to speak, the word "marriage" from your vocabulary. But then, with your last letter you have been upsetting your little J. all over again. —- O mother! You are, of course, right in saying that most people my age are either married or in search of a mate, but let me repeat: there are exceptions. Perhaps you have sensed it all along without daring to admit it to yourself that *I am* such an exception. From your writing I gather that you were misled by my mentioning Virginia's charm, but nobody knows better than she that my affection for her has always been a kind of brotherly love. If God had any purpose in mind when he let me come into this world, he most certainly did not think of me as a husband or father. Your letter somehow compels me to state plainly — once and for all — that I have no intention to be "locked" in wedlock. Science is my mistress, my tyrant, my fate. I would like to be more

explicit, but I am afraid to upset you even more than I already have. Yet, let me tell you that I love you as much as ever, and that you have been — and are — the most wonderful mother I could wish for.
      Your little J.,
    somewhat grown up by now.

***

Diary Entry:
Löbau Castle, August 17, 1539

 Slight depression this morning. No idea why. Long walk. Dr. Tiedemann ready to continue his narration. Must take good notes.

***

To be included in *Vita Cop.*:
Löbau Castle, August 18, 1539.

 When meeting with the bishop I asked him what he remembered about the summer of 1506, when Dr. N. discovered that Jupiter's retrograde motion was an illusion caused by the earth's journey around the sun.
 The bishop's face lit up. "That was his first breakthrough toward the confirmation of his heliocentric universe", he said. "It was perhaps the most triumphant moment of his life, and certainly unforgettable for me. Prior to that day, Nic had been surprisingly reticent about his

progress. I was quietly reading in my study and completely taken by surprise when all of a sudden he rushed in, whirling a set of notes in his left hand, breathlessly panting, 'It's working! I can't believe it! It's working!' Then pounding my shoulder, and bursting again into speech, 'You have to see this. Eureka, eureka! I have to show it to you!'

The concept became soon clear to me, although I did not grasp all the intricacies of the spacial relationships. Yet I was convinced that he had made a momentous discovery, and I said so. He stood there, still shaking with excitement and tears in his eyes.

'The sun is not moving. It's fixed!" he repeated. "It's the earth which moves around the sun. I can't believe it!'

'The earth is moving!' I echoed and tempted by a devilish thought I blurted out, 'You mean to say that the earth has been moving ever since Joshua commanded the sun to stand still!' I hadn't finished the sentence when I regretted my blunder and noticed the change in his mood. A few minutes later he told me that he needed to be alone. As I learned later, he took a walk along the shore of the lagoon. It was late in the afternoon, and when he finally returned, it was almost midnight. He looked pale and haggard, and I could see that doubts gnawed on his conscience. The Church, he argued, could never accept a moving earth. They would condemn his theory and him. I reminded him that the Church had been anxious to correct the inadequacies of Ptolemy's system for many years and that it would support any theory that was scientifically sound. Waving his arms in protest he slumped into a chair, and

The Stranger behind the Copernican Revolution

I realized that he was completely exhausted.

'They *want* reform,' he said, 'but not one that sets the earth in motion and contradicts the Scriptures. The Church will always rely on the Bible as its final authority. It has to.' I agreed in principle but insisted that the Scriptures had always been subject to interpretation.

'In minor ways,' he replied, 'This is more serious than you think - much more so. What I propose is not a shift in viewpoint, it's subversion; it's poison for the faith. During my walk I pondered the consequences, and looked at the problem through the eyes of the Church. There is no question that my findings will be denounced as heresy, and you know what that means. I don't think I am willing to face that."

For the next three days we argued back and forth until he agreed to consult his uncle and possibly some other high-ranking church officials.

What he had not expected was the enthusiasm with which his uncle hailed his nephew's discovery; nor did he foresee the gusto with which Bishop Lucas spread the news. Under these circumstances Nic finally agreed to summarize his ideas in the script that came to your attention, the *Commentariolus*."

***

Diary Entry:
Löbau Castle, August 23, 1539

Up early and depressed again for no good reason. No hint of this to anyone! —- Instead of working on *Narratio,* I am adding details to the notes taken on the terrace. Things are falling into place. — Late afternoon: walk around the park and out of town into the woods. No lift, no high. If Nature is a great healer, she failed my test today. What is the matter with me? No, I am not. No! — Must write to Mother again. Apologize? Cheer her up? With what? The fight over the cat at our arrival? — Better work on *Narratio.*

***

Diary Entry:
Löbau Castle, August 24, 1539

Shortly before noon my Mentor returned, high-spirited and eager to continue the preparation of the manuscript. He looks well, but the cold has affected his hearing! I almost have to shout. Loved his embrace. Wonderful to see him ready for work.

***
Diary Entry:
Löbau Castle, August 26, 1539

Excellent work session with my Mentor, but hoarse from yelling — instead of reading the text. A lingering cold? Some progress on *Narratio Prima*. — Finish that letter?

***
To his Mother,
Thomasina de Porris in Bregenz
Löbau Castle, August 28, 1539

Time certainly has wings in this place. Still comfortably housed in the bishop's residence and regally cared for by two servants, I am, once more, in a state of consummate happiness. When I am not, I am thinking of my last letter to you. Not that I regret anything I said or that I wanted to change a single word. But I wish I could have said it all without hurting you; yet to this day I would not know how! Unfortunately, the slightest waffling on my part lets you dream of wedding bells and five or six bambinos. — Mother, I had to be honest, because the time had come for burdening you with the truth, once and for all. Please forgive me.

Having said this, I feel relieved and in the mood to make good on my promise to let you in on some of my recent experiences. Apart from the qualms regarding my last letter to you, life has been most enjoyable. For more than three weeks I have been celebrating it under the roof of this most remarkable pontiff. As I said before, you would like him, and secretly I relish the thought that your Pastor Sebelius will shudder when you tell him about my fixation on this "son of Sodom"; but that's the pastor's problem. And let me assure you, mother, that you don't have to worry about the immortal soul of your Lutheran J. There are no zealots hiding in the closets of this castle, and believe me, Bishop Tiedemann is a most tolerant man: half of his relatives have turned Protestant while —— hold your breath ——- he remains on the best of terms with them! He even encouraged one of his nephews to study with Melanchthon and Luther at Wittenberg. What do you say to that?

By a strange coincidence, the sister of that nephew arrived here about the same time as we did, and you should have seen the travel outfit and bearing of that Danzig *doña* —- swank and sophisticated as if she had just stepped out of the royal palace in Madrid. I would never have guessed that she was the mother of those two undisciplined youngsters in her coach, a six-year-old boy and a slightly older girl, both totally soiled with dirt from the road. The young lady had decided to "surprise" her uncle and arrived unannounced with her two patrician brats and three incompetent servants in tow. Fortunately they stayed for only two nights, but on arrival they caused a minor calamity that you'll love to hear about.

Pulling into the yard only minutes behind us, and unwilling to listen to the guard, the young lady forged her way upstairs to the bishop's office where we were chatting with our host, admiring his "Lutheran" cats, Martinus and Katie, in whose presence his Excellency had been dictating letters to a scribe when we arrived. We had just sat down, when the door was flung open, and, squealing with excitement, in dashed the niece, hugging and kissing the bishop and enveloping all of us in a cloud of perfume. With raised eyebrows and open mouths my Mentor and I watched this inordinate display of affection, unaware that the boy and his sister were pouncing head over heels on the cats.

While Martinus struggled in vain to escape from the girl's clutching, the boy found himself empty-handed. Fuming with rage about his loss and jealous of his sister's catch, he tried to wrest Martinus from her, pulling ferociously on the cat's tail and hind legs, and finally dragging cat and girl to the floor. Martinus, by now all claws and despair, leaped off to freedom while the young mother, alerted by the screams and alarmed by the oozing of blood, threatened to faint.

"All will be well", declared the bishop, with a calming voice, raising and waving his hands as if he were blessing a crowd. "Why don't we get your nanny", he said, "and then Dr. Nicolaus will see what he remembers from his days as Bishop Lucas's most trusted physician and a miracle doctor in Varmia.

The scribe was instantly at hand with a small kit of medical paraphernalia and had a delightful way of handling the little devils, charming them into sitting side by side so that "all the people in the room could get a good look at what some nasty cats can do to innocent children."

So much for today. But I should add that my monthly allowance (which Dr. Anger insists on calling my maternal pension) is reaching me with amazing regularity and I am, as I have always been, immensely grateful for your generosity. Not that there is much of a chance to spend money at Frauenburg or in Löbau, but it will certainly roll through my fingers once I get to Danzig and Königsberg. Good God, I can hardly wait to see those places —- and be sure that I'll try to write to you from there.

Your impatient son
Joachim.

***

Diary Entry:
Löbau Castle, September 16, 1539

Yesterday Dr. Tiedemann received the books I ordered through Paul. Today the Bishop found the letters and papers that seemed to be lost. Coincident? He still wants to go through them to make sure they are in chronological order. — I can't wait to see them! I can't wait!

***

Diary Entry:
Löbau Castle, September 17, 1539

What do dreams tell us? Anything? Pharaoh's seven cows .... Last night's dream seemed more real than life. I was in Königsberg, presenting Duke Albrecht with our map of the Prussian coastline. He looked much younger than in the woodcut that Dr. Tiedemann showed me. In appearance and demeanor he resembled Melanchthon, trying to cajole me into moving to Königsberg and help him found a university there. Would I be going?

Heinrich Zell was with me, explaining to the Duke his work on the map and pointing out some egregious flaws in the older maps. He did that well, but in the end he embarrassed me by making an ass of himself, getting drunk at the banquet and lambasting the Duke and Spindler. O Henricus! Thank goodness, a dream.

***

Diary Entry:
Löbau Castle, September 18, 1539

Finally Dr. T. brought the papers this morning. Took two days for him to go through them — an eternity.

But what an experience to just hold them, just touch them! He wants everything back, and I understand that. —— Worked all day on them. Exciting material; unfortunately a lot of small scribbling that is hard on my eyes. Some cryptic abbreviations. Will he be able and willing to make them out?

***

To his friend and colleague Paul Eber
in Wittenberg
Löbau Castle, September 20, 1539

    The books you bought for Bishop Tiedemann arrived in record time, and his Excellency was impressed and delighted beyond measure. That you included for me the latest publications of Melanchthon as well as Camerarius' belated tribute to the great Erasmus was a special surprise. I do not know how to thank you sufficiently, because the joy of receiving the books so speedily seems magically related to the reappearance of a real "treasure" which, on closer inspection, made me ecstatic: letters and notes that give us a clear picture of the relationship between Tiedemann Giese and Copernicus, especially before and after my Mentor's return from Italy. In other words, I held in my hands a biographer's dream. Giese had mentioned these papers to me but said that he had not seen them since his move to Löbau.
    Our time at Löbau Castle is gradually coming to an end. My Mentor is not in best shape; he has not quite overcome a cold that affected his hearing, and he is longing to get back to Frauenburg. I am amazed

Upper section of a manuscript page from Copernicus' *The Revolutions of the Heavenly Spheres*

The page shows Copernicus' revision of the Ptolemaic universe: The sun [sol] in the center of the universe and the earth [tellus] with its moon [luna] in the third sphere from the sun — or, beginning the count with the outermost orbit of the "fixed stars", as in the drawing above, the earth is in the fifth orbit.

Signature of Copernicus

that he is still able to work on the revisions with me three and four hours a day. The bishop is trying to keep us here, as he enjoys his company and likes the idea that we work on the script under his roof. If you were to peek in on our "sessions", you would encounter a few anomalies. For one thing, we do not go over the text silently, as you might expect, but your friend Rheticus invariably "volunteers" to read it to him because of his limited eyesight. Bent over the large pages of the manuscript I strain my voice, enunciating distinctly (if not shouting!) each word of my Mentor's heavy Latin prose, watching at the same time for a sign that my message has reached the master. Plowing through the text like that, I have learned to give myself an occasional breather by asking questions which, often enough, I could answer myself. Fortunately, he likes to explain and he does it well. In this way we go on until my voice cracks. That's when we smile at each other, when my Mentor gets up, expresses confidence that he'll recover his hearing, and calls it quits for the day. — It's all a little grotesque at times, but you don't have to feel sorry for me. Enervating as it may be, it's all worth the effort.

Rather upsetting is the latest news from Varmia, where Bishop Dantiscus continues to crusade for virtue and morality. He is now trying to unfrock and exile our friend Alexander Scultetus, an action unheard of since this happened to my Mentor's brother Andreas many years ago. In his prime Scultetus was Varmia's plenipotentiary at the Vatican, where he was quick to acquire the habits which are now causing his woes. Times have changed in the Catholic camp, he tells me, and with his wry humor he blames the "Wittenberg Pope", our revered Dr. Martinus, for his trouble, arguing that Luther's break with Rome forced the Church to change its time-honored tune, forcing the clergy to dance to the beat of an "unnatural morality". If Scultetus were a docile soul, my Mentor tells me, he could still settle his case with the bishop, but he refuses to humble himself before a man with a libertine past. My Mentor and I admire him for that. If exiled, you might want to look out for him in Wittenberg . . .

Take care, my friend, and please tell Melanchthon that Henricus and I are working hard for the glory of his university. No matter what he decides concerning my leave and my status, I am determined to stay up here until Copernicus' book is ready for publication.

***

Diary Entry:
Löbau Castle, September 25, 1539

Pleasant dream about S. —- Long walk with Dr. N. who agreed to have his trigonometry published in Wittenberg. — Our bags are packed. We'll be leaving tomorrow.

The coat of arms of the town of Löbau (Lubawa) shows a bishop with his crosier and a sword. Although Löbau lost the distinction of being the center of a diocese in 1824, modern Lubowa retains the coat of arms in memory of its more glorious past.

***

To his Mother,
Thomasina de Porris in Bregenz
Frauenburg, October 3, 1539

Your latest story about Pastor Sebelius shows me that the man is out of his mind. How can you believe that God is sitting in the *Primum Mobile* on the outskirts of the universe watching every one of us day and night? And he has seen him sitting there? — I hope you'll come to understand that that is just as absurd as his claim that in Heaven Catholics will have to sit on wooden stools, while Lutherans get featherbeds and armchairs. — I was, however, relieved to see that you are distancing yourself from that idea. I understand, of course, that you want to retain the Pastor as a family friend. That he doesn't like me, I don't mind. I sensed that from the moment I shook hands with him, and somehow I relish the thought that he chose theology after his failure in mathematics and cartography.

Much as I cherish the memory of Löbau Castle, I am happy to be back in my Mentor's old study. Here in the tower by the Cathedral I feel at home, and strangely enough, I even savor the Spartan appointments of this abode. It's the place where I do my writing and thinking, and the fact that my Mentor's great book was written here lifts my spirit whenever I think of it. For the final scrutiny of his manuscript, for what he calls his "joy and daily grind", I always meet him at his residence, in *museo suo*. With his hearing more or less

Frauenburg Cathedral

The church where Copernicus worshiped for more than thirty years and where he was buried in 1543. Begun in 1329, the cathedral was dedicated to the Virgin Mary [the Frau].

restored, he is in the best of spirits and enjoys my occasional jokes about your Pastor Sebelius.

Here is an exceptional experience that I must share with you. It happened on the day after our return from Löbau when we enjoyed one of those rare days of autumn, windless and sunny. Late in the afternoon I went up to my favorite lookout high on the rampart from where I have an unrestricted view of the Lagoon, the spit and the Baltic Sea. You would love it! There I sat enjoying the soothing warmth radiating from the bricks of the battlement while the sun gradually moved toward the horizon, pouring shades of orange, red and purple over sky and sea. Drilled in Greek mythology by Dr. Philippus, I saw Helios driving his chariot across the heavens, and thought simultaneously of my Mentor's discovery that this glorious spectacle of the sun "sinking" into the Baltic Sea (or any "setting" sun on earth) is in reality an illusion caused by the earth's rotation; but this novel insight — far from diminishing my enjoyment — did only add to my delight. And then suddenly, while I was still lost in the afterglow, an idea which had vaguely occupied my mind magically presented itself in all its details, setting my mind on fire. It's a project, Mother, larger than life! To complete it will take years and require a whole team of mathematicians as well as sponsors to finance it. But I am not afraid of the obstacles nor of the labor involved. I have discussed it with Dr. Nikolaus, and both of us are convinced that it will fly and usher in a new era in astronomy. The resulting publication will be a huge tome with tables of trigonometric functions which will improve the accuracy of cosmic distances by an unprecedented degree and carry the name of your little J., your slightly cross-grained offspring, to the remotest places on this globe. Something to chew on for our friend Sebelius with his armchairs and featherbeds!

***

To Johannes Schöner,
astronomer and geographer in Nuremberg
Frauenburg, October 5, 1539

    Some unusual news, my friend! It's late at night, but I am still wide awake, excited from the day's events: the arrival of my friend Spindler (on his way from Königsberg to Danzig), our meeting with Copernicus and, above all, an extraordinary moment of insight — each and all of these make this day unforgettable. The "insight" concerns one of those fortuitous instances when the mind of a creative genius is "intuitively drawn toward a new breakthrough". I don't think that I am exaggerating when I say that most of our scientific work, interesting as it may be, is rather pedestrian. Aren't we, after all, just adding bits and pieces of observation to the store of human knowledge? And am I wrong in saying that most of our so-called contributions do not really change things — or change very little? There are days when I find thoughts like that rather depressing. You, on the other hand, would probably

The Stranger behind the Copernican Revolution

argue that we all are preparing the stage for an exceptional genius to surprise or shock us with a stunning, discovery, an unexpected vista of far-reaching promise. I can clearly see that Copernicus has made such a stunning discovery. But how did he do it? What considerations or thought experiments emboldened him to make that contribution? Today, I believe, that I found the answer — or part of the answer — in a bundle of notes, ideas exchanged between him and Tiedemann Giese. To my embarrassment I must confess that I did not grasp their significance when I copied them mechanically at great speed for their possible use in my Mentor's biography. But looking at them from anew together with Spindler, the scales fell from my eyes. Imagine the thrill and amazement when I realized that certain questions they raised were instrumental in opening the door to a new age in science and changing the way we think about ourselves and our place in the universe. What I find so amazing is the *utter simplicity* and the obvious logic of these thoughts. And yet, we had to wait for a man like Copernicus to recognize and pursue their scientific potential. — Let me quote from my notes:

1. Why are the sun and the moon, which have been classified as planets since time immemorial, the only "planets" that do not regress in their respective paths? If planets are known to divert from their courses in predictable loops, why do we count the sun and moon as planets?

2. Does it really make sense to put the glorious sun, the biggest and brightest object in the sky, in the

same class of heavenly bodies as little Mercury or Venus?

3. If the world is a book that God gave us to read intelligently, should we not conclude that He assigned the sun a very special place in his universe, perhaps *the central position,* which so aptly reflects the sun's significance?

When I handed the script to Spindler he read it in silence, sat there for a long time and shook his head. In a husky voice he finally said, "Why did no one else raise these issues before? Some 1400 years have passed since Ptolemy's time, and hundreds of astronomers *could* have asked these same questions. They seem so obvious! And yet, why didn't Peurbach or a genius like Regiomontanus pose and pursue them?"

"Or . . . *we ourselves!*", I blurted out half jokingly, to which Spindler responded wryly, "I suspect that we were either too stupid to recognize their worth, or too conceited to waste our time." And if others *have* tried to tackle these issues in the last thousand years, they certainly did so without an impact". —- I could not agree more. Of course, I am still struck by Spindler's snappy reply, ". . . too stupid or too conceited to waste our time." —- But on the other hand, I am elated by this unusual insight and wish I could see your reaction and hear your comments. — Lacking wings or magic power to transport myself to Nuremberg, I have the good fortune that Spindler will take this letter to Danzig tomorrow so that it will get to Nuremberg in record time.

***
To his Mother,
Thomasina de Porris in Bregenz
Königsberg, October, 21, 1539

Your second-last letter still reached me in Löbau. It was so lovely that I have read it many times; I am still carrying it with me, and your growing interest in my work and my Mentor's discovery makes me happy — happy beyond measure.

For almost a week I have been enjoying myself in Königsberg, an up-and-coming seaport some forty miles north of Frauenburg. The city is less impressive than Danzig or Elbing, but a place with a bright future. The inn of my choice, the Golden Lion, seems to anticipate the projected growth and prosperity of the town. It's remarkable for its comfort and elegance, and outshines all other "lions" in town. I would say that it meets your criteria for gracious living, mother, and it has an extraordinary kitchen. My attraction to such oases of luxury must be an inherited trait, and I thank God (at least once a week!) that He has given me a mother who understands such weakness and whose munificence allows me to indulge in an occasional lift in lifestyle. — *Vivat Lucullus.*

If I had chosen one of the dingy old inns in town, I would never have met a new friend, whom I adore and for whose charm and urbanity you would instantly fall! He's a man in his forties, mother, a rather wealthy merchant from Danzig, who is presently exploring investment opportunities in Königsberg. We happened to share a table for the evening meal on the day of my arrival, and from that time on a magical bond of friendship and brotherly love seems to unite us. As he owns shares in a number of seagoing vessels, he has been actively looking for a solution to the biggest problem in navigation: finding a foolproof method to ascertain the longitudinal position of a ship at sea. You may not know this, but that is primarily an astronomical problem; yet, to my surprise he knew more than I about the subject and has studied Johann Werner's idea to solve the problem with a table of lunar positions. I am just mentioning this, because he wants me to delve into Werner's work in hope of reducing shipwrecks and other costly errors at sea. I love to talk to this man, who has an amazing range of interests, and the hours we spend together are very special to both of us. Needless to say, he has invited me to stay at his house in Danzig whenever I get to that city, and I'll certainly take advantage of his offer.

You may remember that my main purpose of going to Königsberg was to meet the sovereign of this region, Duke Albrecht of Prussia, to whom I intend to dedicate a map that will be printed in Danzig next year. Your Pastor Sebelius might prick up his ears when he hears that. Please tell him that the Duke plans to entice your smart little J. to teach at the new university by making him an irresistible offer. (Brag a little! . . . but brag with moderation!). I love you, mother.

J.

***

To his friend and colleague Paul Eber
in Wittenberg
Königsberg, October 22, 1539

After three days of drizzle, the sun is shining once more on this bustling commercial center at the mouth of the River Pregel, where people are moving toward a more prosperous time than ever in its two or three hundred year history. —- Education ranks high with the Duke, and within four or five years the first university in this region will open its doors. Being a learned man himself, Albrecht has the ambition of making Prussia one of the best educated principalities in Europe. As a staunch supporter of Luther, he questioned me more than two hours about Saxony, the structure of our university, Dr. Martinus and, most of all, about Melanchthon's skill in attracting seasoned scholars to Wittenberg. He is fully aware that he will have a problem luring established scientists to this remote territory; yet this new university will be an excellent opportunity for advancement.

The meeting with him went so well that he asked me to stay for dinner. —- In other words, you may spread the news that I made a big splash at court. Above all, Luther and Melanchthon must hear that Duke Albrecht will be mightily flattered by the dedication and gift of the map. Lay it on! It will be a magnificent piece of cartography created by two illustrious scientists of their university. — That much for public consumption.

Beginning of a thank-you letter by Rheticus for a gift received from Duke Albrecht of Prussia. (August 28, 1541)

Not to be divulged is the following report on the ensuing dinner party, which you would have enjoyed. In the company of a small group of intelligent and lively young courtiers I felt at home from the very beginning. And as you may suspect, our old friend and colleague Spindler was among them, bright and vivacious as ever. — Not that I cared much for the food which, for the most part, consisted of culinary aberrations of Prussian cuisine, among them meatballs larded with pieces of herring. However, there was beer and wine in abundance which let our spirits soar. To my surprise, the Duke began to show a side not seen before: a penchant for prurient humor and scatological jokes, so help me God! Before long, he was ablaze to hear my scoop from Wittenberg, lascivious "dish" which I served with some flair and authority. Finally I topped it

The Stranger behind the Copernican Revolution

all off with your excerpts from Lemnius' *Pornomachia*. — Luther would send me to hell, if he knew that I carry those verses in my back pocket! To this day, I haven't shown them to Copernicus, but here was an ideal audience for them. You should have heard the laughter caused by Lemnius' verse and seen the eyes brimming with tears when I read to them the hilarious passage culminating in Luther's bedroom. — After five months in the somewhat purified atmosphere of Frauenburg and Löbau Castle, I had almost forgotten how much I enjoy this type of socializing. It was a refreshing experience, liberating beyond words! I felt, once more, to be back in the *real* world, where Bacchus dances around the ivory tower and permits us to partake in the oddities and excesses of this extraordinary century, its boisterous merrymaking, its bent for slapstick and goofiness, and its affinity for racy banter and prurient humor. In spirit I was back with you in our circle of friends in Wittenberg. How I wish you could have been part of it all! Amen.

***

To Johannes Schöner,
astronomer and geographer in Nuremberg
Frauenburg, November 20, 1539

You would be ecstatic, my dear Johannes, if you could see the progress I have made with the summary of Copernicus' book. Three weeks of uninterrupted and fruitful work are a wonderful experience! Day after day I was blessed with a power of concentration that I have not known since I left Wittenberg last spring. I can hardly believe it, but I am almost finished, and let me assure you: this little book will make history. A Danzig publisher by the name of Franz Rhode has agreed to print it, and print it soon! The credit for the recommendation goes to an old friend of Copernicus, Bishop Tiedemann Giese, to whom I am also indebted for that invaluable treasure of biographical information from which I quoted to you in my last letter.

I was glad to hear about the progress you are making with your edition of the works of Regiomontanus, and I agree with you that he *was* perhaps the greatest scientific genius of the last century. However, I do not share your view that his untimely death prevented him from discovering what we ought to call from now on the *Copernican Universe*. Nevertheless, I am delighted that I have your permission to dedicate the slim volume of my *Narratio* to you, so that also *your* name will be associated with the unbelievable and most confounding discovery of our age.

***

To Johannes Schöner,
astronomer and geographer in Nuremberg
Danzig, December 18, 1539

I am writing to you from this old Hanseatic seaport, where I am caught in a snowstorm that would be remembered in Nuremberg as the storm of the century. I have not seen anything like it in my whole life, and there seems to be no end in sight for this storm. Strangely enough, there is a bizarre longing in me that wants this spectacle to go on forever.

To let you in on yesterday's most important event: I finally handed the manuscript of my *Narratio* to the printer, and leaving Rhode's house I felt elated. I pictured in my mind's eye the reaction to this publication: acclaim and veneration on one side, and shock and dismay on the other — Europe in jubilation and uproar from Cracow to Madrid! Typesetting will start before the New Year; Heinrich Zell is ready to proofread, and early in March your copies will be delivered to your doorstep.

While the drifting snow outside of my window has reached heights from five to six feet, I am comfortably settled in a heated room. What a joy, what a luxury, one of the benefits of being born late in an age of advancing civilization!

Let me confess that I felt slightly hurt by your well-intentioned question whether my stay in Frauenburg is really worth the effort and the financial sacrifice. I had assumed that in writing to you I had left no doubt that

DE LIBRIS REVO-
LVTIONVM ERVDITISSI-
MI VIRI, ET MATHEMATICI
excellentiff. reuerendi D. Doctoris
Nicolai Copernici Torunnæi Cano
nici Vuarmacienfis, Narratio Prima
ad clariff. Virum D. Ioan. Schone-
rum, per M. Georgium Ioachi-
mum Rheticum, una cum
Encomio Boruffiæ
fcripta.

ALCINOVS.

Δᾶ δ' ἰλινδίριον ἀναι τῇ γνώμῃ τὸν
μάλλοντα φιλοσοφᾶν.

GEORGIVS VOGELINVS ME-
dicus Lectori.

Antiquis ignota Viris, mirandacp noftri
  Temporis ingenijs ifte Libellus habet.
Nam ratione noua ftellarum quæritur ordo,
  Terracp iam currit, credita ftare prius.
Artibus inuentis celebris fit docta Vetuftas,
  Ne modo laus ftudijs defit, honorcp nouis.
Non hoc iudicium metuunt, limamcp periti
  Ingenij, folus liuor obeffe poteft.
At ualeat liuor, paucis etiam ifta probentur·
  Sufficiet, doctis fi placuere Viris.

B A S I L E AE,

Narratio Prima

The first edition of Rheticus' summary of
Copernicus' work (Danzig 1540)
was soon sold out. The second edition appeared
the following year in Basel.

this is an experience I would not want to miss for anything; and if Copernicus lived at one of the remotest outposts in the *New World*, I would be prepared to face the dangers and hardships to reach him and come to know his extraordinary work.

Reading your letter once more, I realize to what extent you are still doubting my Mentor's theory. Basically you are still asking the same question you posed during my visit in Nuremberg, namely, "How can anybody with a sound mind arrive at such conclusion when all visual evidence points toward a stationary earth?" You might remember that, at that time, I could only answer you with a shrug of uncertainty and puzzlement, but today I am convinced that the Copernican theory is scientifically sound. If my *Narratio Prima* should not convince you, his *De Revolutionibus* with all its details certainly will.

I am utterly thrilled when I think that during my stay in this north country I was not only able to bring his theory to the attention of the world, but that in his biography I will be able to pinpoint for posterity the exact moment when our Earth was set in motion, when a lonely scientist on the outskirts of civilization "lifted the anchor" and furnished the first evidence that this globe of ours is no more than a planet orbiting the sun.
—- If you asked me to name my Mentor's greatest feat, I would point at his unbelievable steadfastness, his determination to pursue a goal that seemed to defy all logic. To persist in his effort year after year (knowing all the time that, in the end, one might be wrong!) requires

the courage and daring of an extraordinary genius, one that might come along once in a century.

I know there will always be those who will point to the Scriptures, quoting some inconclusive lines and insisting that only a fool could have come up with such a theory; yet they forget that in the end a fool stands there empty-handed. Copernicus does not! When his book appears, you'll see that quite clearly, and I predict that an irrefutable proof of the earth's motion will come with the next great advance in our science.

The biographical notes I took during my stay at Löbau Castle are like magic stones which allow me to revisit those trying moments in my Mentor's life, those turning points which finally led to his discovery. I can hardly wait for the day when I will be sitting in your study relating to you the remarkable journey of this unusual man.

* *
*

Biographical Comment

Rheticus stayed and worked with Copernicus until, in the summer of 1541, the revisions of *The Revolutions of the Heavenly Spheres* were completed and the manuscript ready for publication. However, when faced with the release of the script, Copernicus stalled once more, all of a sudden unwilling to part with his life's work. We do not know if he acted out of fear that the manuscript might be lost on the way to Nuremberg, or if this was a last attempt to delay the printing, possibly until after his death. In any case, Rheticus, once again, gave proof of his utter determination to see the book published. He offered to copy the voluminous tome for the printer and completed that daunting task in about six exhausting weeks. When he finally left Varmia with his own copy, the original remained in Frauenburg in the hands of its author.

We have no reason to believe that Rheticus intended to stay in Nuremberg to oversee the long, drawn-out process of printing. He soon resumed his teaching duties at Wittenberg and moved to the University of Leipzig in the following year. The responsibility of supervising the printing process he left to Andreas Osiander, a Nuremberg theologian with a mathematical background. While studying the manuscript, Osiander's concern about the controversial aspects of Copernicus' work grew. Apprehensive of the unforeseeable repercussions, and troubled by serious doubts about the validity of the new system, he asked Copernicus to add a brief statement to the effect that his sun-centered universe was merely a hypothesis which facilitated astronomical calculations. A note of this kind, he argued, would appease Catholics and Protestants alike. When Copernicus flatly rejected this proposal, Osiander, fearing for his career and safety, decided to solve the problem single-handedly by composing an unsigned preface, asserting that no claims were made in this book that the heliocentric universe corresponded to the actual structure of the celestial spheres. In this way Osiander protected himself and his publisher, assuring the book a less controversial reception than it would otherwise have had.

There is no evidence that Rheticus was in collusion with Osiander. Keenly aware of Osiander's predicament and cognizant of the ameliorating effect of the preface, he could hardly have been upset. However, a vehement protest came from Tiedemann Giese, who felt that Copernicus, his longtime friend, had been betrayed. Yet, in the end, even Giese may have found some solace in the fact that Copernicus never came to see Osiander's insert. When the great book arrived in Frauenburg in May 1543, Copernicus lay on his deathbed, too weak to scrutinize the *oeuvre* that had been the passion of his life.

* *
*

The Stranger behind the Copernican Revolution

***

To his Mother,
Thomasina de Porris in Bregenz
Leipzig, June 20, 1543

    I can hardly believe that four weeks have passed since we met in Nuremberg, where you expressed the wish to take two copies of the latest edition of my *Narratio* with you to Bregenz. I was touched and delighted by that gesture and in my mind I can see you proudly showing the book to friend and foe.
    The days I spent with you in Nuremberg stand out in my memory, and in my mind I have returned many times to the garden of the Hallers, where I listened to you recounting my father's experiments in alchemy and his commitment to science. That was as much an inspiration to me as your lovely letter some years ago in which you elaborated on his life, the trial and his final execution as a "sorcerer". Let me thank you again for having found the strength to share those experiences with me. Although I remember my father quite well, I knew very little about the trial and the fact that he was beheaded. How wise of you to have taken me at the critical point to your parents in Verona. Did I ever tell you that I read your letter to Copernicus, and also to Bishop Tiedemann? Both men were deeply touched by your account and dismayed that our judicial system would sanction such procedures. A friend of Bishop Tiedemann's, a prominent jurist, who is fighting for revisions of the law, has

been frustrated for years by the snail's pace of judicial changes. It may take a century or more, he told the bishop, to rectify inequities of such magnitude.

I was sorry to hear that you ignored my advice *not* to ask Pastor Sebelius any further questions about Copernicus' theory. To do that is as senseless as asking the devil to lead you to Heaven. It makes perfect sense to me that the pastor would not even look at Copernicus' *Revolutions of the Heavenly Spheres*. He probably wouldn't understand it anyway. After all, it wasn't written for failures in math, nor for religious edification. From his "armchair" in Heaven the pastor might finally get his first view of our moving Earth.

With all my love,
your son J.

***

To his Mother,
Thomasina de Porris in Bregenz
Leipzig, September 22, 1543

Coming home from a long relaxing walk through this enjoyable town, I found your letter from Salzburg! Surprise and joy! And to top it all, the wonderful news of Virginia's wedding! I am delighted — and glad to hear that you enjoyed the festivities and find her husband "acceptable". God bless their marriage!

Regarding Pastor Sebelius, let me honestly say that

The Stranger behind the Copernican Revolution

I have never meant to belittle your friendship with him and his long-standing relationship with your husband. But it makes me sad that he can worry you with some silly remarks about my salvation. I wish you had the courage to stop him whenever he starts with such nonsense. What does Sebelius know about my Mentor (blessed be his soul!) and my relationship to the Almighty? I can assure you that God loves me as much as he loves anyone in pursuit of scientific Truth. As scientists, we are practitioners in the service of God, charged with the mission to unveil the marvels of His creation. Only a nitwit could call my Mentor a "malefactor" and his book "a threat to religion". Far from destroying our faith, it lets us stand in awe of God's ingenious universe.

I am fully aware that Copernicus' book will vex and irk the philistines and the "inflexibles". But what does that matter? With God's help and our ever increasing know-how in science and technology we are moving toward a bright future.

I always marvel at the extraordinary advances that we made in less than a century, and I was thrilled when you pointed out, that your mother was born when the first printed book left Gutenberg's workshop, that you came into this world when Columbus reached the New World, and that I happened to see the light of day when my Mentor made his amazing discovery, a few years before Luther nailed his 95 theses onto the church door in Wittenberg. These are world-shaking events which tell me that God deemed us ready for them — and for more! Rest assured, Mother, my most ambitious undertaking, those tables of trigonometric functions, are bound to become a new landmark in the history of mathematics and will bring us closer to an understanding of our mind-boggling universe.

Since the release of my Mentor's *oeuvre* last March, I have often contemplated its impact on our time and on ages to come. With a certain gratification I am musing about my part in all this. Complete uncertainty lay ahead when I left Wittenberg to reach and persuade my Mentor to publish. At times, things looked outright hopeless, but looking back on it all, it was a task richly rewarding: unforgettable the morning when I packed my copy of his manuscript and at the final farewell he said, "Let's set the Earth in motion!" High on my horse, with tears in my eyes I was riding off, on cloud nine… all the way to Wittenberg — and finally — finally to Nuremberg.

Leipzig continues to be a wonderful place, and I have no regrets about my move. It's a city of considerable wealth with a flourishing culture and an intellectual climate that favors scientific progress. Here your son feels at home, loves his work as much as he loves his colleagues and is likely to stay in this place, a passionate worker as always, and a happy man.

                    Your little J.

* *
*

Epilogue

Eight years later the unforeseen happened. Under the influence of wine and impelled by his sexual preference for men, Rheticus became involved with one of his students, a certain Hans Meusel, whose father brought court charges against him. The consequences were disastrous. Even though the authorities at the University of Leipzig tried to diffuse the matter, Rheticus decided to flee. He was tried *in absentia* and officially banned from the city for ninety-nine years. Rumors about the incident spread quickly and exaggerated accounts became the favorite gossip in university circles north of the Alps. Soberly assessing his professional options, he settled in Prague and decided to go into medicine. When he graduated three years later, he was forty and destitute. However, his medical practice in the Polish capital of Cracow soon flourished. The Polish nobility as well as members of the royal family favored the learned and urbane physician.

In spite of his professional success as a doctor, mathematics and astronomy remained Rheticus' passion. The gigantic project that "set his mind on fire and gave wings to his imagination" during his first year with Copernicus, those trigonometric tables for computing cosmic distances stayed foremost in his mind. With his accumulating wealth and the patronage of Emperor Maximilian II and others, he founded and maintained his own research institute, employing up to six assistants to compute the values of sines, cosines and tangents to the tenth decimal for every tenth degree of an arc.

Toward the end of his life Rheticus saw himself rewarded with the friendship and admiration of a young astronomer who visited him under circumstances amazingly similar to those which brought him to Copernicus some thirty years earlier. According to his own testimony, "history seemed to repeat itself", when young Valentin Otho entered his life, urging him to complete his prodigious research project.

When Rheticus' died in 1574, his huge data bank passed to Valentin Otho, who completed the work and finally, in 1596, published the tables under the title *Opus palatinum de triangulis*. As a tool to calculate cosmic distances with unprecedented accuracy the work was hailed by astronomers and used throughout the 17th century. At the time of its publication, astronomy was on the verge of new exciting discoveries: within the next twenty-three years Johannes Kepler formulated the laws of planetary motion, while Galileo constructed his first telescopes, allowing the human eye to recognize the ruggedness of the lunar landscape and discover the first four of Jupiter's moons. In the decades that followed, more and more circumstantial evidence accumulated to confirm the validity of Copernicus' theory. Yet, almost two hundred years passed after the publication of Copernicus' work until astronomical instruments were refined enough to allow James Bradley in 1728 to furnish the first irrefutable proof of the earth's revolution around the sun.

Opus Palatinum
Title page of Rheticus' trigonometric tables, published posthumously by Valentin Otho in 1596.

Historical Figures:
Biographical Highlights

Albrecht (Albert) von Brandenburg (1490-1568). First Duke of Prussia. Before his conversion to the Lutheran faith, he was Grandmaster of the *Monastic State of Teutonic Knights*. In 1525 he secularized the territory and (with the consent of his uncle, the King of Poland) renamed it the *Duchy of Prussia*. He strongly promoted education in Prussia and founded the University of Königsberg (1544).

Camerarius, Joachim (1500-1574). Professor in Nuremberg, Tübingen and Leipzig; after Erasmus (1466?-1536), one of the most renowned scholars of the 16th century. Among his publications is a first biography of Melanchthon.

Copernicus, Andreas (ca.1471-ca.1523). Older brother of Nicolaus Copernicus. Canon at Frauenburg. Because of his unconventional lifestyle, administrative irregularities and an incurable disease (leprosy?) he was exiled from Varmia. He is believed to have died in Italy.

Copernicus, Nicolaus (1473-1543). Under the guardianship of his uncle, Lucas Watzenrode, he began his studies of mathematics, astronomy and the liberal arts at the University of Cracow (1491-1494) and added – at the universities of Bologna, Padua and Ferrara – canon law, Greek and medicine to his subjects. He received his Doctor of Canon Law degree in 1503 but discontinued the study of medicine without a medical degree. Upon his return to Varmia in 1504 he became his uncle's secretary, advisor and personal physician. As early as 1497, while Copernicus was still a student in Italy, his uncle, as Prince-Bishop of Varmia, had managed to secure for him a canonry at Frauenburg, which meant financial security for life. — With his famous work, *On the Revolution of the Heavenly Spheres,* Copernicus shattered humanity's ancient belief in a stationary earth by providing the first mathematical evidence that the earth was orbiting the sun like a planet.

Dantiscus, Johannes (1485-1548). Humanist name of Johannes Flachsbinder. As a diplomat he served three Polish kings until 1532. While pursuing his diplomatic career in various capitals of Europe, he entered the clergy, managed to receive a Doctor of Law degree from the University of Vienna (1516), got himself crowned Poet Laureate by Emperor Maximilian II, and became Bishop of Kulm in 1530. From 1537 to his death he was Prince-Bishop of Varmia, anxious to raise the moral standard of his clergy. About 12000 letters and other documents are extant.

De Porris, Thomasina (ca.1490-ca.1554). She married the physician Georg Iserin in Feldkirch (Austria), where she gave birth to Rheticus in 1514. After her husband's execution as a sorcerer in 1528, she and her two children were not permitted to retain the family name of

The Stranger behind the Copernican Revolution

Iserin. At that point she decided to use her maiden name, de Porris, for herself and her offspring.

Eber, Paul (1511-1569). Together with his friend Rheticus, he joined the Wittenberg faculty in 1536, eventually becoming a prominent Protestant theologian. Among his publications is a Latin *History of the Jewish people since their return from the Babylonian Exile* (1548).

Ferber, Moritz (Maurice). Bishop of Varmia (1523-1537). Maternal uncle of Copernicus. Under his leadership Varmia recovered quickly from the destruction wreaked by the Teutonic Knights during their war with the Polish Crown.

Giese, Tiedemann (1480-1550). Considered the most learned humanist in northern Poland. A canon at the cathedral of Frauenburg since 1504, he was a colleague of Copernicus for more than thirty years as well as his closest friend. In 1538 Giese was elected Bishop of the neighboring diocese of Culm, and in 1548 he became Prince-Bishop of Varmia. He inherited Copernicus' papers and passionately defended the Copernican idea.

Lemnius, Simon (1507?- 1550). Humanist and Neo-Latin poet. His slanderous epigrams (1538) prompted Luther to call him a "shit-poet" and to banish him from Wittenberg. Lemnius' revenge was another publication with vicious personal attacks on Luther and his wife Catherine (*Monachopornomachia*. Reprint: Munich 1919).

Luther, Katharina (Catherine), neé von Bora (1499-1552). In 1523 she escaped — with Luther's help — from the convent where she had been a nun. After several aborted attempts on Luther's part to find a husband for this dowerless no-nonesense young woman, she indicated that she would be willing to marry him. According to Luther, the union was a happy one. She was an efficient housekeeper and bore him six children.

Luther, Martin (1483-1546). Professor of Theology at the University of Wittenberg, and since the 1520s leader of the Protestant Reformation. His official protest against the sale of indulgences and the adoration of saints (1517) were among the chief causes that led to an unintended break with the Roman Catholic Church. In 1525 he married Katharina (Catherine) von Bora.

Melanchthon. Humanist name of Philipp Schwarzert (1497-1560). Lutheran theologian at the University of Wittenberg; Luther's closest associate and leader of the Protestant movement after Luther's death (1546). Reformer of the German educational system and advocate of religious reconciliation.

Novara, Dominico Maria di (1454-1504). Since 1493 Professor of Astronomy at Bologna. His writings are largely lost. In *De Revolutionibus,* Copernicus mentions observing the occultation of the star Aldebaran by the moon (1497) while being with Novara.

Osiander, Andreas (1498-1552). Nuremberg theologian, important and controversial figure in the history of early Protestantism. Prior to seeing Copernicus' work through the press, he had become experienced in supervising the printing of mathematical texts. In 1549 he was appointed Professor of Theology at the University of Königsberg.

Otho, Lucius Valentin (1550?-1602). A student of mathematics at Wittenberg until August 1573, when he left for Cracow and became Rheticus' assistant. After Rheticus' death (Dec. 1574), Otho continued to work on Rheticus' tables of trigonometric functions, financially supported by Emperor Maximilian II, the Elector of Saxony and, from 1587 onward, by the Counts Palatine, to whom Rheticus' oeuvre, *Opus Palatinum de triangulis* was dedicated in 1596.

Ptolemy, Claudius (ca.100-ca. 160 A.D.) Greek astronomer, geographer and mathematician in Alexandria. He consolidated and amended most of the scientific knowledge of his time. In his handbook on astronomy, the *Almagest,* he perfected the model of Aristotle's and Hipparchus' earth-centered universe; in addition, he catalogued the position of more than a thousand "fixed stars", and managed to explain, with the help of epicycles and other devices, the irregular motions of the planets. The *Almagest* remained the standard textbook for astronomers until the seventeenth century.

Regiomontanus. Humanist name of Johannes Müller (1436-1476). Generally considered the most gifted mathematician and astronomer of his century. Some of his works were still used by Copernicus and Galileo. His premature death during a trip to Rome came as a shock to his contemporaries.

Rheticus. Humanist name of Georg Joachim von Lauchen (1514-1574). After his father was beheaded as a sorcerer (1528), his wife and children were no longer permitted to carry their family name. His mother reverted to her maiden name of *de Porris*, while Rheticus, feeling more German than Italian, translated her name into German as *von Lauchen*. — In the history of astronomy Rheticus is remembered as the person to whom science owes the publication of Copernicus' famous work, *On the Revolution of the Heavenly Spheres*. As a mathematician in his own right he is, most of all, credited with the creation of a huge database of trigonometric functions which vastly improved astronomical measurements. While financing and managing his own research institute at Cracow, he declined offers of professorships from the universities of Vienna and Paris.

Schillings, Anna. A relative of Copernicus who for a number of years became his housekeeper. Reports or rumors about a possible liaison between her and Copernicus led Bishop Dantiscus to press for her dismissal in December 1538.

Schöner, Johannes (1477-1547). Professor of mathematics in Nuremberg. He designed many of the oldest globes showing the newly discovered Americas. In addition to his own publications, he edited the works of Regiomontanus and Johannes Werner. Rheticus dedicated his summary of Copernicus' work to him (*Narratio Prima.* Danzig 1540).

Scultetus (also Sculteti), Alexander (ca. 1480- ?). Canon and historiographer at Frauenburg. Earlier in his career he had acted as Varmia's representative at the papal court. A longtime associate of Copernicus, who was particularly close to him after Tiedemann Giese's departure from Frauenburg (1537). His "Chronology of the World" (Chronographia sive annales) ended with the year 1545 and was posthumously published in Rome (1595).

Von der Trenck, Achatius (ca. 1497-.?) Younger contemporary of Copernicus, in charge of the episcopal estates around Allenstein at the time of Rheticus' visit. He belonged to the Prussian nobility whose members, along with patrician families of the Hanseatic cities like Danzig and Thorn, dominated in the upper echelons of the Church in Royal Prussia and Varmia during the fifteenth and sixteenth centuries.

Watzenrode (also Watzelrode), Lucas (1447-1512). Maternal uncle of Copernicus. Doctor of Canon Law (University of Bologna, 1473). As Prince-Bishop of Varmia

(1489-1512), he secured canonries for his two nephews, Andreas and Nicolaus Copernicus, enabling the latter to work freely on his theory of a heliocentric universe.

Werner, Johannes (1468-1528). Nuremberg mathematician and cartographer. One of the few mathematicians of his time to venture into the field of stereography and spherical geometry.

Zell, Heinrich (ca.1520-1564). His map of Prussia, published in 1542, became his first scientific contribution. Other cartographic publications followed. In 1555 he became the librarian of Duke Albrecht of Prussia in Königsberg. The catalogue of the University of Königsberg of that year lists him as "excellent mathematician and cartographer". It seems that his collaboration with Rheticus ended when his assistantship was terminated after the completion of the map of Prussia.

* *
*